SMITH

Glinška Ulica 10
Vic
61000 Ljubljana
Yugoslavia

061 - 262-028

KINEMATIC HYDROLOGY
AND MODELLING

OTHER TITLES IN THIS SERIES

KINEMATIC HYDROLOGY AND MODELLING

DAVID STEPHENSON

Department of Civil Engineering, University of the Witwatersrand, 1 Jan Smuts Avenue, 2001 Johannesburg, South Africa

and

MICHAEL E. MEADOWS

Department of Civil Engineering, University of South Carolina, Columbia, SC 29208, U.S.A.

ELSEVIER

Amsterdam — Oxford — New York — Tokyo   1986

ELSEVIER SCIENCE PUBLISHERS B.V.
Sara Burgerhartstraat 25
P.O. Box 211, 1000 AE Amsterdam, The Netherlands

Distributors for the United States and Canada:

ELSEVIER SCIENCE PUBLISHING COMPANY INC.
52, Vanderbilt Avenue
New York, NY 10017, U.S.A.

Library of Congress Cataloging-in-Publication Data

Stephenson, David, 1943–
    Kinematic hydrology and modelling.

    (Developments in water science ; 26)
    Bibliography: p.
    Includes indexes.
    1. Runoff--Mathematical models. 2. Groundwater
flow--Mathematical models. I. Meadows, Michael E.
II. Title. III. Series.
GB980.S74 1986   551.48'8'0724   86-2175
ISBN 0-444-42616-7

ISBN 0-444-42616-7 (Vol. 26)
ISBN 0-444-41669-2 (Series)

Printed in The Netherlands

PREFACE

Many stormwater design engineers and indeed hydrologists will be frustrated by the lack of hydraulic principles in some of the conventional methods of flood calculation. The Rational method and unit hydrograph methods are easy to apply but limited in accuracy and versatility. Kinematic hydrology is the next logical step in sophistication before the full hydrodynamic equations are resorted to. The kinematic equations in fact comprise the continuity equation and a hydraulic resistance equation. In many cases solution of these equations for flow rates and water depths is simple and explicit. In more complicated problems the equations may be used to simulate the runoff process.

Unfortunately much of the literature on the kinematic method has been highly mathematical and often of an experimental nature. The equations, graphs and models published are therefore of little use to the practical engineer, and may discourage him from using this method. In fact once confidence is gained, the method can be applied in simple form to a variety of catchments. The term kinematic refers to movement where accelerations are negligible – which is generally applicable to overland and shallow stream flow.

The book is aimed at both the theoretician and the practitioner. Thus the mathematical sections are useful if modelling is required, but the chapters on design charts could be read with very little mathematical understanding other than a basic appreciation of the kinematic method. Little mathematical background is required, and no computer knowledge is necessary for those sections. It is hoped that the peak flow charts will provide an alternative to the Rational method and the SCS method for estimating runoff. Similarly the dimensionless hydrographs are competitive with unit hydrograph methods. The user will gradually become aware of the fact that the kinematic method is fairly easy to apply if simple solutions are required. It also permits consideration of many more factors than some other methods of flood calculation, which in turn can only improve accuracy and provide for greater understanding of the runoff process.

Of course the kinematic method is not the final answer in hydrology. There are many questions still to be answered, and some degree of simplification is still required. Although the method provides a logical way of visualizing runoff, actual runoff from many catchments comprises part overland, subsurface and interface flow. The combined effect cannot easily be modelled. Also water does not run off rural catchments in a sheet – it frequently forms rivulets and is diverted by obstacles which can be loosely termed roughness. Some of these factors can be accounted for by adjusting the hydraulic factors used in the equations, or calibrating models.

Results of research and development are now advanced and experience in application is required before general acceptance of the kinematic method can be hoped for. In particular the ability to select soil losses, roughnesses and catchment geometry to adequately describe the hydraulics of the system, can only be gained with experience.

The scope of the kinematic method is therefore unlimited from the point of view of the researcher with an enquiring mind. Some of the theoretical considerations are taken further in chapter 2 on kinematic equations, 4 on assumptions and 5 on numerical theory for modelling.

On the other hand the practitioner is probably more interested in the best answer available. He may manage quite sufficiently reading only chapter 3 on peak flows, chapter 6 with dimensionless hydrographs and possibly chapter 7 on marginal effects and 9 with some examples of the value of the techniques. Hopefully he will be inspired to go into modelling, which may bring in chapter 8 on flow in conduits, and 10, 11 and 12 with examples of computer models of various catchments.

Much of the material in this book is derived from notes for a course presented by the authors. There is copious reference to previous research in kinematic hydrology, as well as new material arising from research by both authors. In particular the senior author was the recipient of a research contract in urban hydrology from the Water Research Commission.

The manuscript was typed into its final form by Janet Robertson, for which the authors are most grateful.

CONTENTS

CHAPTER 1.   INTRODUCTION

CHAPTER 2.   ANALYSIS OF RUNOFF

CHAPTER 3.   HYDROGRAPH SHAPE AND PEAK FLOWS

CHAPTER 4.   KINEMATIC ASSUMPTIONS

CHAPTER 8. CONDUIT FLOW

CHAPTER 9. URBAN CATCHMENT MANAGEMENT

CHAPTER 10. KINEMATIC MODELLING

CHAPTER 11. APPLICATIONS OF KINEMATIC MODELLING

CHAPTER 12. GROUNDWATER FLOW

CHAPTER 1

INTRODUCTION

HISTORICAL REVIEW

Kinematic hydrology provides a method for estimating stormwater runoff rates and volumes. It is particularly useful for flood calculation. It is a relatively new term embracing techniques which have been applied for many decades. Kinematic hydrology is decidedly more hydraulically correct than some of the more common methods of flood estimation such as the rational method, time-area methods, the Soil Conservation Service (SCS) method and unit hydrograph methods. The kinematic method is based on the continuity equation and a flow resistance equation, both basic hydraulic equations.

It was the American hydrologist, Horton, (generally associated with infiltration) who in 1934 carried out the earliest recorded scientific studies of overland flow. Later Keulegan (1945) applied the continuity and momentum equations conjunctively for overland flow analysis. He investigated the magnitude of the various terms in the dynamic equation of St. Venant and indicated that a simplified form of the equation, now termed the kinematic equation, would be adequate for overland flow.

An in-depth analysis of the differential continuity and resistance equations was undertaken by Lighthill and Whitham (1955) to whom the designation kinematic waves can be attributed. They also first studied the phenomenon of kinematic shock which can be applied to discontinuities in flow and water depth. Although they suggested the kinematic approach for overland flow modelling, it was Henderson and Wooding (1964) who obtained analytical solutions to the kinematic wave equations for simple plane and channel shapes. A generalization of the catchment stream model was also described by Eagleson (1967).

The full dynamic equations for one-dimensional incompressible flow in open channels were set down by St. Venant in 1871. These equations were for gradually varied unsteady flow such as flood waves. The idea of graphical integration using characteristic lines was first suggested by Massau in 1889. On the other hand Greco and Panattoni (1977) indicate that implicit solution by finite differences is the most efficient method by computer, avoiding instability and giving rapid convergence. Various numerical methods of solution of the kinematic equations were investigated by Kibler and Woolhiser (1970). The step length in finite difference schemes plays an important role in the stability of the solution (Singh,

1977). Non-convergence was investigated for plane cascades by Croley and Hunt (1981). Brakensiek (1966) used numerical solutions to the kinematic wave equations for the analysis of surface runoff from rural watersheds. He probably did not realise the extent to which numerical modelling would advance in later years using the kinematic equation and square x-t grids. The latter approach does not warrant appendage of the term 'wave' to kinematic since discontinuities are lost in the simplified numerical method.

Wooding (1965 and 1966) presented a comprehensive review of the theory of kinematic waves and used numerical solutions to derive equations for the rising and falling limbs of hydrographs for simple planes and channel configurations. During the 1970's the equations were applied to more complicated catchment shapes (Schaake, 1975), in particular the catchment-stream model, the converging catchment and cascades of planes. Although analytical solutions are available for some cases the majority of solutions are numerical, and dimensionless hydrographs facilitate the use of the results of the studies (Constantinides and Stephenson, 1982). Since the studies by Henderson and Wooding (1964) and Iwagaki (1955) the shock wave phenomenon has not really received much attention and for this reason the use of the name kinematic theory is now considered adequate as it implies a more general applicability than to waves. In fact Borah and Prasad (1982) indicate shock waves may in fact not exist in some cases where predicted using the kinematic equations. This is because the kinematic equations may not apply where the spatial variation in depth is large. Even the St. Venant equations may not suffice to describe rapid varied flow, as vertical accelerations are not considered.

Woolhiser and Liggett (1967) investigated the applicability of the kinematic equations and proposed a dimensionless parameter indicating whether the equations are adequate for any particular case with simple geometry. More recent research (Morris and Woolhiser, 1980) has investigated in greater detail the applicability of the kinematic equations to different conditions.

The application of kinematic theory has more recently been extended to problems such as dynamic storms (Stephenson, 1984a), detention storage (Stephenson, 1984b), urban drainage networks (Green, 1984) and to the effects of urbanization and storm runoff (Stephenson, 1983).

There is as yet little general data available on surface water losses (infiltration, and retention) to be used with kinematic equations. Skaggs (1982) reviewed infiltration mechanics including the popular Horton model and more advanced Green-Ampt model.

The majority of papers differentiate between surface and subsurface flow, i.e. overland flow is treated independently. Rovey et al. (1977) developed an interactive infiltration model to account for non-uniform soil losses. A further development by Freeze (1972) allows for contributions from re-appearing shallow groundwater flow in a saturated aquifer.

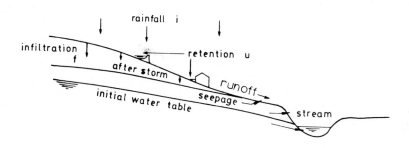

Fig. 1.1   Simplified catchment longitudinal section

Definitions

Some terms used in this text are used in different context elsewhere so to avoid confusion particularly with respect to times, some definitions are given below.

Time to equilibrium (t_m) is the time taken from the commencement of precipitation until the water profile down the catchment is in equilibrium and inflow equals outflow everywhere, i.e. runoff rate is equal to excess rainfall rate, assuming steady precipitation and losses.

Time of concentration (t_c) is the time from the commencement of precipitation until the effect of excess precipitation everywhere in the catchment has appeared at the outlet. It is equal to the time to equilibrium for steady excess rain using kinematic theory whereas it is equal to travel time with time area-theory. It is demonstrated later that for a simple plane, kinematic theory yields

$$t_c = (L i_e^{1-m} / \alpha)^{1/m} \tag{1.1}$$

where L is the length of flow path, and flow velocity $V = \alpha y^{m-1}$ where y is water depth and m and α are coefficients defined by the equation $q = \alpha y^m$ where q is the flow rate per unit width.

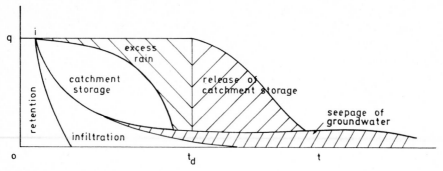

Fig. 1.2   Catchment water balance

Travel time (t_t) is the time for a particle of water to proceed from the most remote part of the catchment to the discharge point. For a plane it is not equal to time of concentration according to kinematic theory since water moves slower than a hydraulic response which travels at wave speed.   It is shown later that for a plane

$$t_c = t_t/m \qquad (1.2)$$

Lag time (t_L) is the time between 50% of precipitation and 50% of runoff. It will be shown that for a plane

$$t_L = mt_c/(1+m) \qquad (1.3)$$

Storm duration t_d is the time from commencement of precipitation until it ceases. Frequently when storm records are analyzed for intensity–duration relationships storm duration is defined as the time during which average storm intensity is a specified figure, so that storms within storms can occur.

Time of excess runoff (t_e) is the time measured from the commencement of runoff. It is therefore less than the time t from the commencement of precipitation by $t_u = u/i$ where u is initial abstraction and i is the precipitation rate (see Fig. 3.3 on page 49).

Units of time are generally seconds if the System International (S.I.) units of metres, seconds and kilograms, or the old English system of foot, seconds and pounds are adopted. Later herein modifications for more practical units e.g. rainfall in mm/h or inches per hour, are introduced.

CLASSICAL HYDROLOGY

For various reasons flood hydrology has been a fairly static subject for many decades. The rational method which was invented over 100 years ago, and hydrograph theory, developed over 50 years ago, are still used extensively. If we reconsider the assumptions and limitations behind these methods we may be prepared to consider developing new techniques more appropriate to our technology and more accurate.

The simple linear hydrology methods were probably developed for ease of manual calculation, and as many hydrologists do not have a strong mathematical background. It is true that some of the standard methods have been programmed for computers. This facilitates the subdivision of catchments but does not eliminate the limitations of many of the assumptions behind the methods.

The current availabilities of computers to all should considerably ease the next step – breaking away from simple input–output methods and introducing more sophisticated hydraulic equations in their stead. It is possible to simulate water flow and water surface profiles with considerable accuracy with the aid of computers, even micro computers. There are various levels of sophistication which can be adopted to suit the problem and the machine available.

These methods are based on solution of finite difference versions of the differential equations of flow. Computations proceed in increments of time at selected intervals in space. There have been numerous advances in numerical methods in mathematics in parallel with the developments in computers. On the other hand the approximation of differentials by finite increments can lead to inaccuracies unless certain rules are complied with. Some of the common problems are instability, numerical diffusion or accumulating errors. The correct finite increments can be selected to approximate the differentials to a first order, second order or greater order if necessary. There are also methods for solving implicit equations such as by gradient convergence or successive approximation. Where a number of simultaneous equations have to be solved over a grid there are matrix methods and relaxation methods available.

One of the greatest aids to the engineer nowadays may be the desk top micro computer. Whereas practitioners tend to shy away from main frame computers (if they can access one at all) the problems of job control language, queing batch jobs, formal programming and debugging and risk of runaway costs are no longer of concern. The kinematic method is intermediate level technology applicable to micro computer solutions, whether analytical solutions or numerical modelling is contemplated.

The basis for much of our hydrology probably originated with an Irish engineer, Mulvaney, in 1851. He proposed an equation for runoff, $Q = KA$. K allows for a rainfall intensity but this was not a significant variable in Britain. The method was taken a step further by introducing an equation for excess rainfall intensity, e.g. the Birmingham formula,

$$i = \frac{40}{20+t} \tag{1.4}$$

where i is in inches per hour and t is the storm duration in minutes. No allowance is made for extreme storms and this equation is for a 1 to 2 year frequency storm. The 20 was accepted by some as representing a time of entry in minutes (equivalent to the defined concentration time of overland flow).

It was assumed that 100% runoff occurred from impermeable areas and none from pervious areas. This assumption was not acceptable in areas of high rainfall intensity and in the United States where Kuichling in 1889 modified the runoff equation to Q = CiA where the coefficient C is a function of the catchment.

The coefficient C is most strongly associated with the average permeability of the catchment – thus 100% runoff would occur if C is unity and no runoff for a completely permeable catchment. Modifications to C are made to account for catchment slope, vegetation cover and so on by various people. It has also been realized that antecedent moisture conditions and severity of the storm (represented by the recurrence interval) can affect C. For instance Rossmiller (1980) proposed the following empirical equation for C:

$$C = 7.7 \times 10^{-7} C_N^3 R^{.5}(.01 C_N)^{-6S^{.2}} (.001 C_N)^{1.48(.15-I)} \left(\frac{M+1}{2}\right)^{.7} \tag{1.5}$$

where R is the recurrence interval, S is bed slope in percent, I is rainfall intensity in inches per hour, M is the fraction of watershed which is impervious and C_N the Soil Conservation Service (SCS) curve number.

The assumption of a unique 'C' for any catchment can lead to significant errors and underestimation of flood runoff. This is demonstrated by Figure 1.3. The runoff rate per unit area for case 'a' is Ci_1. If the same C is used for case b, where a higher rainfall intensity occurs, the loss will be greater and the runoff proportional. A loss which is independent of rainfall intensity however would produce a runoff as for case c, which is proportionally greater than for case b. The assumption for case b thus results in an underestimate of flood runoff.

In general then, it is implied in the Rational method that runoff intensity is linearly proportional to rainfall intensity. This also assumes that the catchment has reached an equilibrium, so it became necessary to estimate the 'concentration time' of catchments. Lloyd-Davies developed this idea in 1905 and proposed that the maximum peak runoff from a catchment occurred for a storm with a duration equal to the concentration time of the catchment. A common equation used for concentration time is

$$t_c = (0.87L^3/H)^{0.385} \qquad\qquad (1.6)$$

where t_c is in hours, L is the length of catchment in km and H the drop in metres, or

$$t_c = (11.6L^3/H)^{0.385} \qquad\qquad (1.7)$$

where L is in miles and H in ft.

The rational method does not produce a complete hydrograph capable of routing and so unit hydrograph theory was developed. The theory was based on the assumption that two units of excess rain produce a hydrograph with ordinates twice those of a hydrograph produced by one unit of excess rain in the same time. The term linear hydrology is often applied to this theory. The time scale is also incremented linearly. Two successive units of rain are assumed to produce two unit hydrographs in succession which can be added together at all points in time. We thus have the S-curve hydrograph which is caused by an infinitely long storm.

Unit hydrographs do not account for the non-linear response of a catchment to excess rain. Neither is the concentration time of any catchment area a unique time, it depends on the flow rate, as seen for instance, in the Manning equation (2.47). In any case the travel time is not the same as the reaction time which is also a function of flow rate. Non linear hydrograph theory on the other hand has met with limited response.

To some extent the error in assuming the travel time is the concentration time is nullified by assuming a full conduit for computation of travel time. The upstream conduits flow at a lower rate than those downstream. When the design storm is occurring for a downstream conduit, upstream conduits will be flowing at less than design capacity as the storm duration will be greater than the design storm for the upper conduits. Thus the assumption of a higher flow and velocity than will occur makes the resulting rate of concentration more nearly that of the true hydrodynamic system.

Another misconception is that the full catchment must contribute for the maximum runoff rate. Besides odd shaped catchments which can by analysed using the tangent method (Watkins, 1962) a true analysis would show many catchments do not contribute from the farthest extremity at peak flow. This is not shown up by the rational method which invariably assumes the entire catchment contributes. It can be demonstrated only if soil-dependent losses are assumed, not rain-dependent losses (e.g. 'C'). It is shown in chapter 3 that if loss is independent of rainfall then a shorter duration storm in many cases produce a greater runoff rate than one which is of duration equal to the time to equilibrium.

rainfall and runoff rates
per unit area of catchment

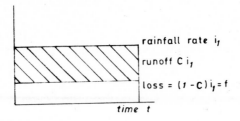

(a) Medium storm

rainfall runoff rate

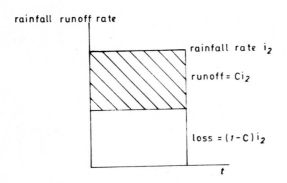

(b) Intense storm assuming same C as in (a) above

rainfall runoff rate

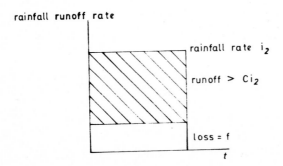

(c) Intense storm with same loss as (a)

Fig. 1.3   Effect of constant C on runoff

HYDRODYNAMIC EQUATIONS

The Navier-Stokes equations for incompressible fluid flow in three dimensions are

$$\rho(\frac{\partial u}{\partial t}+u\frac{\partial u}{\partial x}+v\frac{\partial u}{\partial y}+w\frac{\partial u}{\partial z}) = X - \frac{\partial p}{\partial x} + \mu(\frac{\partial^2 u}{\partial x^2}+\frac{\partial^2 u}{\partial y^2}+\frac{\partial^2 u}{\partial z^2}) \tag{1.8}$$

$$\rho(\frac{\partial v}{\partial t}+u\frac{\partial v}{\partial x}+v\frac{\partial v}{\partial y}+w\frac{\partial v}{\partial z}) = Y - \frac{\partial p}{\partial y} + \mu(\frac{\partial^2 v}{\partial x^2}+\frac{\partial^2 v}{\partial y^2}+\frac{\partial^2 v}{\partial z^2}) \tag{1.9}$$

$$\rho(\frac{\partial w}{\partial t}+u\frac{\partial w}{\partial x}+v\frac{\partial w}{\partial y}+w\frac{\partial w}{\partial z}) = Z - \frac{\partial p}{\partial z} + \mu(\frac{\partial^2 w}{\partial x^2}+\frac{\partial^2 w}{\partial y^2}+\frac{\partial^2 w}{\partial z^2}) \tag{1.10}$$

where ρ is the mass density of the fluid, u,v,w, are the velocity components in the x,y,z directions respectively, X,Y,Z, are the body forces per unit volume, p is the pressure and μ is viscosity. In addition to these three dynamic equations we have the continuity equation

$$\frac{\partial u}{\partial x} + \frac{\partial v}{\partial y} + \frac{\partial w}{\partial z} = 0 \tag{1.11}$$

Although these four equations theoretically describe flow in any situation, from the point of view of civil and hydraulic engineers they suffer a number of drawbacks. For instance viscous forces should be replaced by turbulent momentum transfer or even by a semi-empirical friction drag equation, e.g. by Manning or Darcy.

It is generally possible to work in one dimension in civil engineering hydraulics. Then the Navier-Stokes equations can be replaced by the St. Venant equations, which also comprise a dynamic equation and a continuity equation, namely

$$\frac{1}{g}\frac{\partial v}{\partial t} + \frac{v}{g}\frac{\partial v}{\partial x} + \frac{\partial y}{\partial x} + S_f - S_o = 0 \tag{1.12}$$

and $\frac{\partial Q}{\partial x} + B\frac{\partial y}{\partial t} = 0$ $\tag{1.13}$

where S_o is the bed slope (positive down in the x direction), S_f is the energy gradient, Q is the flow rate, B the surface width, A the cross sectional area and P the wetted perimeter. It will be seen on close inspection that the St. Venant equations are similar in many terms to the Navier-Stokes equations.

The solution of the St. Venant equation is, however, a difficult enough task for the hydrologist or civil engineer. The classical solution is by the method of characteristics which can easily be portrayed graphically. Computer solution of the equation in various forms is now more common. Rapid solution of a finite difference form of the St. Venant equations in a simplified form can easily be undertaken on, for instance, micro computers.

For the majority of overland flow cases and in many channel and conduit flow situations the St. Venant equations can be replaced by the following two equations (see chapter 2).

$i_e = \dfrac{input}{Area.}$

Continuity $\qquad \dfrac{\partial Q}{\partial x} + B\dfrac{\partial y}{\partial t} = i_e$ $\qquad\qquad$ (1.14)

Dynamics $S_o = S_f$ $\qquad\qquad\qquad\qquad\qquad\qquad\qquad$ (1.15)

where i_e is the input per unit area of surface (e.g. excess rainfall intensity).

These equations are termed the kinematic equations. Equation (1.15) merely states that the bed slope can be substituted for the energy gradient in a friction equation.

For overland sheet flow q per unit width these equations become

$\dfrac{\partial q}{\partial x} + \dfrac{\partial y}{\partial t} = i_e$ $\qquad\qquad\qquad\qquad\qquad\qquad$ (1.16)

$q = \alpha y^m$ $\qquad\qquad\qquad\qquad\qquad\qquad\qquad\qquad$ (1.17)

where i_e is the excess rainfall rate.

It is further a simple matter to transform the kinematic equations (1.14) and (1.15) into equations applicable to storage reservoirs with interlinking conduits;

$\Delta Q + A\dfrac{\partial h}{\partial t} = q$ $\qquad\qquad\qquad\qquad\qquad\qquad$ (1.18)

and $\quad \Delta H/L = KQ^m$ $\qquad\qquad\qquad\qquad\qquad\qquad$ (1.19)

Here A is the reservoir surface area, Q is the net inflow from connecting pipes and q is the drawoff from a reservoir with water level h. The second equation is applicable to closed conduits and in fact is simpler than the open channel kinematic equation since the variable flow depth is eliminated.

When the common node between conduits is an open reservoir the continuity equation will predict the rate of change in water level. If the conduits or pipes connect at a closed node it is necessary to solve simultaneously for head at the node and flow in the connecting pipes.

Many methods are available for this, but the linear method (Stephenson, 1984b) is particularly suitable. That procedure requires minimal data preparation and solution is faster than the manual node iterative correction procedure of Hardy Cross because it is implicit, that is heads of all nodes are solved for simultaneously. The kinematic method of continuous simulation is a versatile technique for analysis of urban storm drainage and water supply pipe networks particularly when operation of storage reservoirs is involved.

The limiting assumptions behind the kinematic method should however be recalled. Although the assumption that the x-differential terms in the dynamic equation is zero is certainly valid, the time differential terms may in some cases not be zero. This effect is magnified by introducing closed conduits with unvarying cross-sectional area. Pressure rises due to change in flow rate can be large, giving rise to water hammer.

In such situations, i.e. when rapid fluctuations in flow are possible, an alternative method of analysis, namely elastic analysis, must be employed. To analyse a network using the water hammer equations involves simultaneously solving the characteristics and continuity equation at each node. Aspects of friction damping require particular attention with this method. In particular the ratio of friction head loss to water hammer head can have an important effect on the speed of solution. When the analyst is only concerned with steady state heads and flows he can artificially speed convergence by suppressing the wave speed i.e. reducing the numerical value used in the computations.

The analyst is thus altering the fit of the mathematical model to the real system. There are approximations and consequently scope for adjustment at a number of stages in the modelling. The following stages are related by the analyst:

Real system (conduits and reservoirs)

Imagined system (what can be visualized)

Mathematical model (differential equations)

Numerical model (finite differences)

Computer model (successive equations)

By adjusting the imagined system one is able to speed convergence of the solution. The finite differences have to be limited according to the Courant criterion (1956) and particularly when friction is involved, another criterion proposed by Wiley (1970)

$$\Delta t \; < \; (\Delta x/c) \; (1 - Sg\Delta t/2v)^{1/2} \hspace{3cm} (1.20)$$

Equation (1.20) indicates that friction affects the stability of numerical solutions. This is however due to the numerical approximation in solving the equations explicitly rather than an instability caused by friction. Friction has generally an important role in kinematic theory. It relates water depth to flow rate i.e. it provides the link between the continuity equation and the hydrograph. Although friction energy loss relationships are well known for stream flow which is fully turbulent and sub-surface flow which is laminar, the process of overland flow is not fully appreciated. Flow depths are small and the dimensions of roughness are comparable with the flow depth. There are complicating influences such as tortuous flow paths around and over boulders, vegetation, structures and other surface disturbances. Rain drops are reported to cause turbulence at lower Reynolds numbers than for conduit flow. Overton and Meadows (1976) indicate turbulent flow persists for sheet flow if the Reynolds number in terms of precipitation rate, $iL/\nu = 20$ to 2000 where i is the precipitation rate (m/s), L is the overland flow path length and ν is the kinematic viscosity of the liquid (water). This would indicate that the energy gradient is proportional to flow rate to the power of $m = 5/3$ if the Manning equation (2.47) is assumed together with the 1/6 power law for velocity distribution. Horton (1938) on the other hand found m was approximately 2 on natural surfaces implying nearly laminar conditions for uniform flow (constant depth in the direction of flow). Actually m = 3 for pure laminar flow.

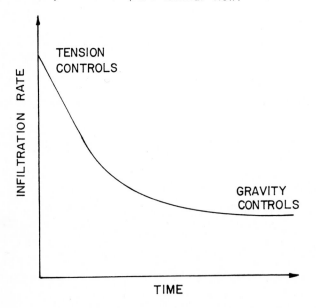

Fig. 1.4 Typical field infiltration curve

INFILTRATION

A major component of a stormwater model is the routine to determine the rainfall excess. Abstractions or losses are subtracted from input rainfall resulting in the rainfall excess which must be routed to the basin outlet.

The losses which must be abstracted from rainfall are:

1. Interception-rainfall caught by vegetation prior to reaching the ground. The amount caught is a function of (a) the species, age, and density of vegetation, (b) character of the storm, and (c) the season of the year. It has been estimated that in a rural watershed as much as 10 to 20 percent of the rainfall during the growing season is intercepted and returned to the atmosphere by evaporation.

2. Depression storage-water caught in small surface pockets and voids held there until it infiltrates or evaporates.

3. Evaporation-water returned to the atmosphere through vaporization. Evaporation is most important when it is not raining; it is negligible during rainfall events when a representative rate is 0.05 mm/hr (0.002 in/hr) (Overton and Meadows, 1976).

4. Infiltration-water lost to the soil. Typically, infiltration is the major abstraction during a rainfall event. Three distinct processes are involved: (a) the movement of water into the soil across the air-soil interface (infiltration); (b) the movement of water through the soil under the influence of gravity and soil suction (percolation); and (c) the depletion of the available volume within the soil (storage depletion).

There are two basic approaches to modelling rainfall excess. Each loss can be modelled separately and the models linked together, or a single model can be developed that lumps the important losses together, usually into infiltration. This latter approach is often followed in event simulation models. Kinematic stormwater models are mostly event models; therefore, we are mostly concerned with infiltration models for the rainfall abstraction model.

A typical field infiltration curve is shown in Figure 1.4. Infiltration begins at an initial high rate and decreases with time to a steady final rate. The forces influencing the movement of water into and through the soil are suction and gravity. During the early stages, the upper soil layer is "thirsty" and infiltration is dominated by suction. With time, the upper centimetre, more or less, of the soil surface becomes saturated and the infiltration rate reduces to that rate at which water moves through the saturated soil. At this point, gravity dominates. As long as the rainfall rate exceeds the instantaneous infiltration rate, or water is ponded on the surface, infiltration will continue at the maximum possible rate, defined by Horton (1933) as the capacity infiltration rate. The effect of rainfall rate on the infiltration curve is next examined. Three general cases for infiltration during a steady rainfall were proposed by Mein and Larson (1973):

Case A: $i < k_s$. (The rainfall rate, i, is less than the saturated soil hydraulic conductivity, k_s.) Under this condition, runoff will not occur, regardless of rainfall duration, because all rainfall will infiltrate.

Case B: $k_s < i < f_p$. (The rainfall rate is less than the capacity infiltration rate, f_p, but is greater than the saturated hydraulic conductivity.) For a short duration rainfall, where i remains less than f_p, all the rain infiltrates. But for a rainfall of long duration, the infiltration capacity will decrease until it equals i, and surface ponding occurs.

Case C: $k_s < f_p < i$. (The rainfall rate is greater than the infiltration capacity.) Under this condition, runoff occurs.

Cases B and C can be considered as two distinct cases; however, infiltration often occurs as a two-phase process combining the two cases. Bodman and Colman (1943) evaluated soil water distribution during infiltration into a uniform, relatively dry soil under surface ponding conditions and established that the typical profile can be divided into four zones as shown in Figure 1.5. The uppermost zone is the saturation zone and varies little in thickness, regardless of the total depth of infiltration. Immediately below this zone, there is a zone of rapid decrease in the water content, which Bodman and Colman called the transition zone; and below it, there occurs a zone of nearly constant moisture called the transmitting zone. This zone increases in length in direct proportion to the volume of infiltrated water. Next, there is the wetting zone which moves downward with a constant shape as infiltration continues. The wetting zone ends at the wetting front, which is the boundary between water penetration and soil at the initial moisture content.

Soil Physics Models

There are two approaches to modelling the infiltration process, soil physics models and hydrologic models. Soil physics models are deterministic models based on the physics of soil moisture movement, while hydrologic models are conceptual and are based on a die-away rate until the final steady rate is reached. The advantage of soil physics models is that the parameters are understood and are measurable; the disadvantage is that soil physics models typically require a large amount of data, including site measures of soil porosity, hydraulic conductivity, soil layering, etc. In comparison, hydrologic models generally have fewer parameters, require less data and are easier to solve; however, the parameters are not alway physically interpretable and cannot be measured, hence they must be established by calibration. A further criticism of hydrologic models is that they oversimplify the infiltration process, particularly during periods of unsteady rain and rainfall less than the soil saturated hydraulic conductivity.

The governing equations for infiltration are the conservation of mass and an equation of motion.

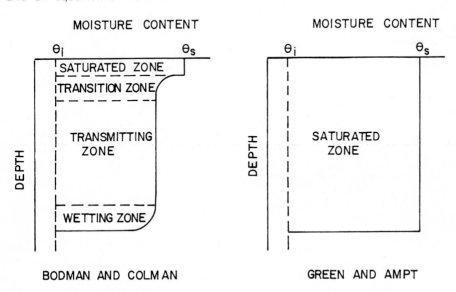

Fig. 1.5 Comparison of Green and Ampt soil moisture profile with Bodman-Colman profile

The conservation of mass equation is

$$\frac{\partial v}{\partial z} + \frac{\partial \theta}{\partial t} = 0 \qquad (1.21)$$

where v is the specific discharge (velocity) vertically, θ is the volumetric moisture content.

The equation of motion is based on Darcy's law for a saturated, homogeneous soil,

$$v = - k \frac{dh}{dz} \qquad (1.22)$$

where v is velocity as defined previously, k is hydraulic conductivity, h is the hydraulic head, dh is the change in head in the direction of flow over the length dz. The negative sign indicates flow is in the direction of decreasing head.

Darcy's law can be generalized to unsaturated flow by expressing the hydraulic head as a function of soil tension or suction, and gravity. During the initial stages of infiltration when the water content is low, the tension force is much larger than the gravity force and the flow process is controlled by tension. As the pores fill, tension is reduced and gravity becomes important. The hydraulic head is then equal to tension ψ plus gravity, z.

$$h = \psi + z \qquad (1.23)$$

and Darcy's law as applied to unsaturated flow is

$$v = - k \left(\frac{\partial \psi}{\partial z} + 1 \right) \qquad (1.24)$$

By combining Eqs. 1.23 and 1.24, we get the governing equation for one-dimensional, vertical, unsaturated flow, known as Richard's equation.

$$\frac{\partial \theta}{\partial t} = \frac{\partial}{z} \left[k \left(\frac{\partial \psi}{\partial z} + 1 \right) \right] \qquad (1.25)$$

where k and ψ are both functions of θ. Due to the nonlinear relationship between hydraulic conductivity, suction and soil moisture, there is no general analytical solution to Eq. 1.25.

The problem is further complicated by hysteresis in that the relationship between suction and moisture content is not unique and single valued. The relationship depends on whether the soil is wetting (infiltration is occurring) or drying (drainage is occurring). These relationships are shown in Figure 1.6. Generally, for a given water content, suction is lower during wetting than during drainage and minor hysteretic loops can occur between the main hysteretic loops. The hysteretic effect is attributed to (1) geometric nonuniformity of individual pores, (2) variations in contact angle in wetting and drainage,

(3) entrapped air, and (4) swelling (Hillel, 1971). Conductivity likewise exhibits a hysteretic effect.

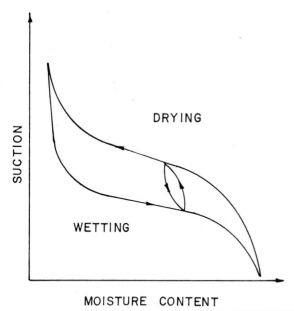

Fig. 1.6   Typical soil suction – moisture relation

Green and Ampt Model

A conceptual model utilizing Darcy's law was proposed by Green and Ampt (1911). Many studies, including those by Mein and Larson (1973), have demonstrated the usefulness of the Green and Ampt model for modelling infiltration. As methods for measuring the model parameters are made easier, it can be expected the model will be more widely applied.

Darcy's law can be written as

$$v = \frac{f}{n} = k(h + L_f + \psi_f)/L_f \tag{1.26}$$

where f is the infiltration rate and v is equal to the vertical velocity, h is the surface ponding depth, L_f is the depth to the wetting front, and ψ_f is suction at the wetting front.

Several assumptions were necessary to write Darcy's law in the form of Eq. 1.26, namely:

1. There exists a distinct and precisely definable wetting front.

2. Suction at the wetting front, ψ_f, remains essentially constant, regardless of time and depth.

3. Above (behind) the wetting front, the soil is uniformly wet and of constant hydraulic conductivity k.

4. Below (in front of) the wetting front, the soil moisture content is relatively unchanged from its initial moisture content, Θ_i.

These assumptions, when checked against the actual soil moisture profile of Bodman and Colman illustrate the approximate nature of the Green and Ampt model. This is shown in Figure 1.5.

The accumulated infiltration depth, F, can be obtained by integrating Eq. 1.26.

$$f = dF/dt = k(h + L_f + \psi_f)/L_f \tag{1.27}$$

or more directly from

$$F = (\Theta_s - \Theta_i)L_f = \Delta\Theta L_f \tag{1.28}$$

where Θ_s is the saturated moisture content and Θ_i is the initial moisture content. The measure of moisture content, Θ, is a volumetric measure, therefore $\Delta\Theta$ is calculated with the relationship

$$\Delta\Theta = (\Theta_s - \Theta_i) = \phi(1 - S_i) \tag{1.29}$$

where ϕ is the soil porosity and S_i is the initial degree of saturation. Applying the relationships in Eqs. 1.28 and 1.29 to Eq. 1.27 and integrating to obtain F gives

$$k_t = F - (\psi_f\Delta\Theta)\ln[1 + F/(\psi_f\Delta\Theta + h\Delta\Theta)] \tag{1.30}$$

which is a nonlinear equation implicit in F and t. An explicit formulation to solve for the incremental infiltration volume during an incremental time interval, Δt, is obtained by rewriting Eq. 1.30. This gives

$$\Delta F = \frac{k\Delta t - 2F_t}{2} + \sqrt{\left(\frac{2F_t - k\Delta t}{2}\right)^2 + 2k\Delta t(h_t\Delta\Theta + \psi_f\Delta\Theta + F_t)} \tag{1.31}$$

where ΔF is the increment in total infiltration from time t to time $t+\Delta t$, and F_t and h_t are the infiltration and ponded depth, respectively, at time t. Therefore, the total infiltration after the time increment is

$$F_{t+\Delta t} = F_t + \Delta F; \quad \text{if } \Delta F < i\Delta t + h_t \tag{1.32a}$$

or

$$F_{t+\Delta t} = F_t + i\Delta t + h_t; \quad \text{if } \Delta F > i\Delta t + h_t \tag{1.32b}$$

where i is the rainfall intensity. If $\Delta F < i\Delta t$ for a time step then excess intensity, i_e, occurs.

$$i_e = i - f = i - \frac{\Delta F}{\Delta t} \tag{1.33}$$

The incremental cumulative infiltration equation, Eq. 1.32, was developed assuming uniform soil properties. However, it can be applied to layered soils, assuming each layer has uniform properties. The required soil properties, i.e. K_s, ψ_f, ϕ, and S_i, and the thickness, d, must be known for each layer. After computing the infiltration during each time interval, the cumulative infiltration volume, F, is compared with the storage capacity of uppermost layer not yet saturated. Once a layer becomes saturated, the infiltration rate is controlled by the conditions in that layer or the next lower layer, whichever gives the smaller rate.

Bouwer (1966) defined the Green and Ampt parameter k to be "the actual hydraulic conductivity in the wetted zone," which is less than the saturated hydraulic conductivity, k_s. He concluded, based on previous work, that k may be taken as about $0.5k_s$. The saturated hydraulic conductivity can be determined by several standard laboratory tests.

Effective saturation is defined as

$$S_e = \frac{\Theta - \Theta_r}{\phi - \Theta_r} \qquad (1.34)$$

where Θ_r is the residual moisture content. Brooks and Corey (1966) observed a straight line relationship

$$S_e = (\psi_b/\psi_c)^{1/B}; \text{ for } \psi_c > \psi_b \qquad (1.35)$$

where ψ_c is capillary pressure head (suction) at a given soil moisture content, Θ; ψ_b is termed bubbling pressure and is defined at the intercept of a straight line plot of effective saturation and capillary pressure head; and B is an index of the pore size distribution. Porous media composed of single grain material have primary porosity (porosity consisting only of spaces between the grains) and tend to have small values of B. Media having secondary porosity (pore spaces also available for flow with aggregates) have large values (>1.0).

The wetting front suction is estimated using the following relationship

$$f = \frac{\eta}{\eta - 1} \frac{\psi_b}{2} \qquad (1.36)$$

where $\eta = 2+3/B$

Hydrologic Infiltration Models

Horton (1939) proposed an infiltration equation to represent the typical infiltration curves observed in double-ring infiltrometer tests. In these experiments, the water is continuously ponded above the soil; therefore, the supply is not limiting and infiltration proceeds at the maximum potential rate. He observed that the infiltration rate was initially high and decreased in time to a steady final rate. The die-away followed a negative exponential very closely. His equation is

$$f = f_c + (f_o - f_c)e^{-kt} \qquad\qquad (1.37)$$

where f is the capacity infiltration rate at time t, f_o and f_c are the initial and final infiltration rates, and k is the infiltration constant which is allegedly a function of soil and vegetation. In theory this equation assumes the air-soil interface is saturated at all times. In practical terms this means that it is assumed the rainfall rate is always greater than infiltration capacity rates, and hence some ponding will always result. This is a major disadvantage in the use of Horton's model since natural rainfall rates are highly variable and therefore frequently fall below the capacity rates. This may not be a problem with high intensity design rainfalls or rainfalls distributed in time to always exceed the capacity infiltration rates.

Holton (1961) proposed a conceptual model of infiltration backed by substantial field experimentation. He recognized from soil physics as the pores fill, the infiltration rate dies away and approaches a steady final rate. The final rate of infiltration f_c was associated with the gravity force at field capacity (and is assumed to equal the soil saturated hydraulic conductivity, k_s). He then formulated a model to relate capacity infiltration rate to the available soil moisture storage volume remaining at any time, F_p, as

$$f = aF_p^n + f_c \qquad\qquad (1.38)$$

The parameters a and n were determined experimentally from infiltrometer plot data. The exponent was found to be about 1.4 for all plots studied and the coefficients varied from 0.2 to 0.8 for the soil-cover complexes studied.

REFERENCES

Beven, K., Dec. 1982. On subsurface stormflow. Predictions with simple kinematic theory for saturated and unsaturated flows. Water Resources Res. 18 (6) pp 1627-33.

Bodman, G.B. and Colman, E.A. 1943. Moisture and energy conditions during downward entry of water into soils. Proc. Soil Science Soc. of America, Vol. 7, pp 116-122.

Borah, D.K. and Prasad, S.N., 1982. Shock structure in kinematic wave routing. In Rainfall-Runoff Relationships, Edt. Singh, V.P., Water Resources Publications, Colorado, 582 pp.

Bouwer, H. 1966. Rapid field measurement of air entry value and hydraulic conductivity of soil as significant parameters in flow system analysis. Water Resources Research, Vol. 2, No. 4, pp 729-738.

Brakensiek, D.L., 1966. Hydrodynamics of overland flow and non-prismatic channels. Trans. ASAE 9 (1), pp 119-122.

Brooks, R.H. and Coley, A.T. 1966. Properties of porous media affecting fluid flow. Journal of the Irrigation and Drainage Division, ASCE, Vol. 92, No. IR2, pp 61-88.

Constantinides, C.A. and Stephenson, D., 1982. Dimensionless hydrographs using kinematic theory, Report 5/1982. Water Systems Research Programme, University of the Witwatersrand, Johannesburg.

Courant, R., Friedrichs, K. and Lewy, H., 1956. On the partial difference equations of mathematical physics. N.Y. Univ. Inst. Maths.

Croley, T.E. and Hunt, B., 1981. Multiple valued and non-convergent solutions in kinematic cascade models, J. Hydrol., 49, pp 121-138.

Dunne, T., 1978. Field studies of hillslope flow processes. Ch. 7, Hillslope Hydrology, Ed. Kirkby, M.J., John Wiley, N.Y.

Eagleson, P., 1967. A distributed linear model for peak catchment discharge. Intl. Hydrol. Symp., Colorado State Univ., Fort Collins, pp 1-18.

Freeze, R.A., 1972. Role of subsurface flow in generating surface runoff. 2, Upstream source areas. Water Resources Research, 8 (5), pp 1272-1283.

Gallati, M. and Maione, U., 1977. Perspective on mathematical models of flood routing, in Mathematical Models for Surface Water Hydrology, Edt. Ciriani, T.A., Maione, U. and Wallis, J.R., Wiley Interscience, 423 pp.

Greco, F. and Panattani, L., 1977. Numerical solution methods of the St. Venant equations. In Mathematical Models for Surface Water Hydrology, Edt. Ciriani, T.A., Maione, U. and Wallis, J.R., Wiley Interscience, 423 pp.

Green, I.R.A., 1984. WITWAT stormwater drainage program. Report 1/1984, Water Systems Research Programme, University of the Witwatersrand, Johannesburg.

Green, W.H. and Ampt, G.A. 1911. Studies of soil physics, 1. The flow of air and water through soils. J. of Agriculture Science, Vol. 4, No. 1, pp 1-24

Henderson, F.M. and Wooding, R.A., 1964. Overland flow and groundwater flow from steady rainfall of finite duration. J. Geophys. Res. 69 (8) pp 1531-1539.

Hillel, D. 1971. Soil and water-physical principles and processes, Academic Press

Holton, H.N. 1961. A concept of infiltration estimates in watershed engineering, U.S. Dept. of Agriculture, Agric. Research Service, No. 41-51, Washington, D.C.

Horton, R.E. 1933. The role of infiltration in the hydrologic cycle. Trans. of the American Geophysical Union, Hydrology Papers, pp 446-460

Horton, R.E., 1938. The interpretation and application of runoff plot experiments with reference to soil erosion problems. Proc. Soil Sci. Soc. Am. 3, pp 340-349.

Horton, R.E. 1939. Approach toward a physical interpretation of infiltation capacity. Proc. Soil Science Soc. of America , Vol. 5, pp 399-417.

Horton, R.E., Leach, H.R., and Van Vliet, R., 1934, Laminar sheet flow. Amer. Geophys. Union, Trans., Part II, pp 393-404.

Iwagaki, Y., 1955. Fundamental studies on the runoff analysis by characteristics. Disaster Prevention Research Institute, Bulletin 10, Kyoto Univ. 25 pp.

Keulegan, G.H., 1945. Spatially varied discharge over a sloping plane. Amer. Geophys. Union Trans. Part 6, pp 956-959.

Kibler, D.F. and Woolhiser, D.A., 1970. The kinematic cascade as a hydrological model. Colorado State Univ. paper 39, Fort Collins, 25 pp.

Kouwen, N., Li, R.M. and Simons, D.B., 1980. Flow resistance in vegetated waterways. Manuscript, Colorado State University, Fort Collins.

Lighthill, F.R.S. and Whitham, G.B., 1955. On kinematic waves, I, Flood measurements in long rivers. Proc. Royal Soc. of London, A, 229, pp 281-316.

Lloyd-Davies, D.E., 1905. The elimination of storm water from sewerage systems. Min. Proc. Instn. Civil Engnrs., 164(2) pp 41-67.

Massau, J., 1889. L'intégration graphique. Assoc. Ingenieurs Sortis des Ecoles Spéciales des Gard, Annales. 435 pp.

Mein, R.G. and Larson, C.L. 1973. Modeling infiltration during a steady rain. Water Resources Research, Vol. 9, No. 2, pp 384-394.

Morris, E.M. and Woolhiser, D.A., 1980. Unsteady one-dimensional flow over a plane: Partial equilibrium and recession hydrographs. Water Resources Research, 16 (2), pp 355-360.

Overton, D.E. and Meadows, M.E., 1976. Stormwater modelling, Academic Press, 358 pp.

Rossmiller, R.L., 1980. The Rational formula revisited. Proc. Intl. Symp. Storm Runoff, Univ. of Kentucky, Lexington.

Rovey, E.W., Woolhiser, D.A. and Smith, R.E., 1977. A distributed kinematic model of upland watersheds. Hydrology Paper 93, Colorado State Univ., Fort Collins, 52 pp.

Schaake, J.C., 1975. Surface waters. Review of geophysics and space physics 13 (13) pp 445-451.

Singh, V.P., 1977. Criterion to choose step length for some numerical methods used in hydrology. J. Hydrol., 33, pp 287-299.

Skaggs, R.W., 1982. Infiltration. Ch. 4, Hydrological Modelling of Small Watersheds, Edt. Haan, C.T., Johnson, H.P. and Brakensiek, D.L., ASAE Monog. 5.

SCS (Soil Conservation Service) 1972. National Engineering Handbook, Secn. 4. Hydrology, Washington, D.C.

Stephenson, D., 1983. The effects of urbanization. Course on Modern Stormwater Drainage Practice, SAICE, Cape Town.

Stephenson, D., 1984a. Kinematic study of effects of storm dynamics of runoff hydrographs. Water S.A. Vol. 10, No.4, Oct. 1984. pp 189-196.

Stephenson, D., 1984b. Kinematic analysis for detention storage. EPA/Users group meeting. Detroit.

Watkins, L.H., 1962. The Design of Urban Sewer Systems, Road Research Techn. paper 55, HMSO, London.

Wooding, R.A., 1965a. A hydraulic model for the catchment-stream problem 1, Kinematic wave theory. J. Hydrology, 3. pp 254-267.

Wooding, R.A., 1965b. A hydraulic model for the catchment-stream problem, II, Numerical solutions, J. Hydrol. 3. pp 268-282.

Wooding, R.A., 1966. A hydraulic model for the catchment-stream problem III, Comparison with runoff observations, J. Hydrology, 4, pp 21-37.

Woolhiser, D.A. and Ligget, J.A. 1967. Unsteady one-dimensional flow over a plane - The rising hydrograph. Water Resources Research, 3 (3), pp 753-771.

Wylie, E.B., 1980. Unsteady free surface flow computations. Proc. ASCE, 96 (HY 11) pp 2241-2251.

CHAPTER 2

ANALYSIS OF RUNOFF

INTRODUCTION

In this chapter a simplified description of the rainfall – runoff mechanism is presented, i.e. one which can be described in equation form. The concept of mass balance whereby input equals outflow plus change in storage, is applied to simple catchments. The build-up of water depth over the catchment when a storm occurs is described as well as the mechanism whereby runoff occurs. The relationship between water depth and flow rate forms an important part in the prediction of flow so the equation of motion (in fact only a simple flow resistance equation in the case of kinematic flow) is introduced.

This simple analysis is confined to a rectangular plane catchment sloping uniformly down in the direction of flow, and flow is assumed overland. The equations of continuity and flow are thus particularly simple. Nevertheless the origin of and the assumptions behind the simplifications are presented. A simple demonstration of the applicability of the kinematic equations is also given. Later other components of flow e.g. sub-surface flow (Beven, 1982) and a more practical assessment of the contribution to streamflow are introduced with modelling. The different-iation of surface and subsurface flow is often more complicated than assumed here (Dunne, 1978).

DYNAMIC EQUATIONS

The equations governing unsteady, one-dimensional overland and open channel flow are derived by applying the principles of conservation of mass and momentum to elemental fluid control volumes. One-dimensional equations actually describe the change in streamflow in two dimensions: vertical and longitudinal. They are classified as one-dimensional since only one spatial variable occurs as an independent variable.

The important assumptions are:

1. The water surface profile varies gradually, which is equivalent to stating the pressure distribution is hydrostatic, i.e., vertical accelerations are small;

2. Resistance to flow can be approximated by steady flow formulae;

3.   The velocity distribution across the wetted area can be represented
     with the cross-sectional average velocity;
4.   Momentum carried to the streamflow from lateral inflow is negligible;
     and
5.   The slope of the channel is small.

     In addition, for this derivation, the channel is assumed rectangular.
This simplifies the mathematics involved and has little effect on the final
form of the governing equations.

Conservation of Mass

     The continuity principle states that the net mass inflow to a control
volume must equal the rate of change of mass stored within the control
volume. Consider the elemental fluid volume shown in Figure 2.1, where Q
is the volumetric flowrate in m^3/s or cfs, q_i is the lateral inflow rate in
m^3/s per m or cfs per foot length of channel, y and A are depth and cross
sectional area of flow in metres and square metres (feet and square
feet), respectively, θ is the slope of the channel with respect to the

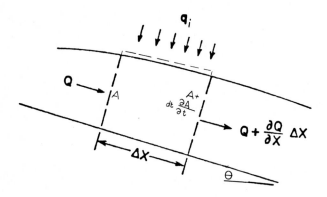

Fig.   2.1 Derivation of continuity equation

horizontal measured as an angle, and x and t are the space and time
coordinates in metres (feet) and seconds. The total inflow to the section is

$$\text{Inflow} = Q + q_i \Delta x \tag{2.1}$$

and the total outflow is

$$\text{Outflow} = Q + \frac{\partial Q}{\partial x} \Delta x \tag{2.2}$$

The change in volume stored in the section is equal to the change in cross-sectional area of flow multiplied by the length of the section.

Change in volume stored $= \dfrac{\partial A}{\partial t}\Delta x$ (2.3)

Combining these quantities according to the above stated principle, dividing by Δx, and rearranging, yields the continuity equation

$$\frac{\partial Q}{\partial x} + \frac{\partial A}{\partial t} = q_i$$ (2.4)

Conservation of Momentum

This second equation is given by Newton's second law of motion which states that the rate of change of momentum is equal to the applied forces. The applied forces, as seen in Figure 2.2, are (1) pressure, (2) gravity, and (3) resistive frictional forces.

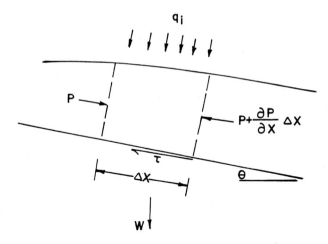

Fig. 2.2   Derivation of momentum equation

Consider forces in the downstream direction as positive. The pressure downslope acts opposite to the pressure upslope and upon summing, the net pressure force becomes

$-\rho gA(\partial y/\partial x)\,\Delta x$

where ρ is the mass density of water and g is the gravitational acceleration.

Similarly, it can be shown that the gravity or weight force acting on the volume of water in the section is given by $\rho gA\Delta x\,\tan\theta$

where, for gradually varied flow, $\tan \theta$ closely corresponds to the channel slope, S_o, and may be expressed as such. This is called the small slope approximation.

Finally, the friction force acting to retard the flow is expressed in terms of an average shear stress

$$- \tau P \Delta x$$

where τ is shear stress and P is wetted perimeter. From the relationship formed by equating head (energy) loss to the work done by the shear force we know that $\tau = \gamma R S_f$, where S_f is the friction slope and γ is the unit weight of liquid. Substituting for τ, and recalling that R=A/P, the following expression for the friction force is obtained.

$$- \gamma R S_f P \Delta x = - \gamma S_f A \Delta x \tag{2.5}$$

The resultant force on the fluid volume in the direction of flow is the summation of the three applied external forces.

$$\rho g A \Delta x \left[-(\partial y/\partial x) + S_o - S_f \right]$$

The change in momentum consists of two parts, the local or temporal momentum change and the spatial or convective momentum change. The local momentum of the fluid is $\rho A \Delta x \, v$, and the local change is just the time derivative

$$\frac{\partial}{\partial t}(\rho A \Delta x v) = \rho \Delta x \left(A \frac{\partial v}{\partial t} + v \frac{\partial A}{\partial t} \right) \tag{2.6}$$

The spatial change in momentum is the rate of momentum change across the control surface. The momentum flux through the control surface is $\rho v^2 A$, and the spatial change is the x-derivative

$$\frac{\partial}{\partial x} (\rho v^2 A) \Delta x = \left(2 A v \frac{\partial v}{\partial x} + v^2 \frac{\partial A}{\partial x} \right) \Delta x \rho \tag{2.7}$$

The total momentum change is the sum of the temporal and spatial momentum changes.

$$\Delta x \rho \left(A \frac{\partial v}{\partial t} + v \frac{\partial A}{\partial t} \right) + v \Delta x \rho \left(v \frac{\partial A}{\partial x} + 2A \frac{\partial v}{\partial x} \right)$$

Substituting the following equivalence from continuity

$$A \frac{\partial v}{\partial x} + v \frac{\partial A}{\partial x} = q_i - \frac{\partial A}{\partial t} \tag{2.8}$$

allows the rate of momentum change to be written as

$$A \Delta x \rho \left(\frac{\partial v}{\partial t} + v \frac{\partial v}{\partial x} + \frac{v q}{A} i \right)$$

Equating this expression with the summation of external forces gives the following familiar form for the conservation of momentum equation.

$$\frac{\partial v}{\partial t} + v \frac{\partial v}{\partial x} + g \frac{\partial y}{\partial x} = g (S_o - S_f) - \frac{v q}{A} i \tag{2.9}$$

where S_o is bed slope, S_f is friction slope and R is hydraulic radius and is equal to A/P.

Eqs. 2.4 and 2.9 can be made applicable to any cross section for both overland and open channel flow, though strictly they apply to rectangular channels only in the present form.

These equations are nonlinear, hyperbolic, partial differential equations and represent a nonlinear, deterministic, distributed, time variant system. They are sometimes referred to as the St. Venant equations.

SIMPLIFIED EQUATIONS

Equations 2.4 and 2.9 are accepted as fully descriptive of one dimensional overland and open channel flow routing. These equations describe both the forward or downstream wave propagation characteristics as well as the backward or upstream characteristics. It is assumed that flood waves in streams move downstream and since hillslope runoff is always downhill, the backward characteristics are simply backwater effects, and in some flow routing instances, they can have substantial impact and control on the flow. As such, these equations are known generally as the *dynamic* wave equations. As a runoff hydrograph passes through a channel reach, the combined effects of channel irregularity, pool and riffle patterns, natural and manmade roughness and gravity forces act to reduce the hydrograph peak assuming lateral inflow is insignificant while lengthening the time base. That is, the peak of the hydrograph is attenuated while the shape is dispersed in time (also in space). The dynamic wave equations account well for hydrograph attenuation. However, two drawbacks to the wholesale general use of these equations are the large data requirements and the necessity for numerical integration. Very often, based on channel geometry and alignment and flood wave characteristics, it is possible to make valid simplifying assumptions that allow one to utilize approximations to the dynamic wave equations. When this is possible, advantages in terms of ease of solution and data requirements are often realized.

Two approximations that have found wide application in engineering practice are the *diffusion* and *kinematic* wave models. The diffusion wave model assumes that the inertia terms in the equation of motion, Eq. 2.9, are negligible compared with the pressure, friction, and gravity terms. Thus, the diffusion model equations are continuity, Eq. 2.4, and the following simplified form of the conservation of momentum equation.

$$\frac{\partial y}{\partial x} = S_o - S_f \qquad\qquad (2.10)$$

For prismatic channels, Eqs. 2.4 and 2.10 are often combined into the single equation

$$\frac{\partial Q}{\partial t} + c \frac{\partial Q}{\partial x} = D \frac{\partial^2 Q}{\partial x^2} \tag{2.11}$$

where c is the wave celerity in m/s (fps) and D is a hydrograph dispersion coefficient in m^2/sec (ft^2/sec). Because Eq. 2.11 is of the form of the classical advection-diffusion equation, it is commonly called the diffusion wave model.

The kinematic model further assumes the pressure term is negligible, reducing Eq. 2.10 to

$$S_o = S_f \tag{2.12}$$

which means the equation of motion can be approximated by a uniform flow formula of the general form

$$Q = ay^b \tag{2.13}$$

where a,b are constants.

Although approximations, both the diffusion and kinematic models have been shown to be fairly good descriptions of the physical phenomemona in a variety of open channel and overland flow routing cases. The kinematic model has been successfully applied to overland flow, to small streams draining upland watersheds, and to slow-rising flood waves. This latter case occurs both in major streams such as the Mississippi River when long duration flood hydrographs resulting from, as an example, spring snowmelt in the U.S. Midwest and Canada, and in small streams where the streamflow hydrograph results principally from lateral stormwater inflow.

THE KINEMATIC EQUATIONS

For overland flow and in many channel flow situations, some of the terms in the dynamic equation (2.9) are insignificant. Neglecting the q_i component one can write the equation as

$$S_f = S_o - \frac{\partial y}{\partial x} - \frac{v}{g} \frac{\partial v}{\partial x} - \frac{1}{g} \frac{\partial v}{\partial t} \tag{2.14}$$

(1)   (2)   (3)   (4)   (5)

The order of magnitude of each of the five terms is evaluated below for a shallow stream. If the bed slope (2) is 0.01, the longitudinal rate of change of water depth (3) is unlikely to exceed 0.1m/100m = 0.001. The longitudinal velocity gradient term (4) will also be less than $(1m/s/10m/s^2)(1m/s/100m)$ = 0.001, and the time rate of change in velocity term (5) will in all probability be less than (1/10)(1/100s) = 0.001.

Terms (3), (4) and (5) are therefore at least an order of magnitude less than (2) for depths up to 1m, and for flow depths less than 0.1m they will be two orders of magnitude less. Those terms can therefore be neglected for the majority of overland flow problems. The inaccuracy in solutions omitting these terms for runoff hydrographs was evaluated by various researchers:

Woolhiser and Liggett (1967) investigated the accuracy of the kinematic approximation and found it to be very good if the dimensionless parameter for planes $S_oL/y_LF_L^2$ is greater than 20 and reasonable if greater than 10. y_L is the depth at the lower end of the plane of length L and slope S_o and F_L is the Froude number $V_L/(gy_L)^{\frac{1}{2}}$. i.e. gS_oL/V_L^2 > 10. Morris and Woolhiser (1980) and Woolhiser (1981) later found the additional criterion $S_oL/y_L >5$ is also required.

The resulting simplified dynamic equation omitting terms (3), (4) and (5) simply states that the friction gradient is equal to the bed gradient. The friction gradient can be evaluated using any suitable friction equation, e.g. that of Manning. The two equations referred to as the kinematic equations are thus the continuity equation which per unit width of overland flow becomes

$$\frac{\partial y}{\partial t} + \frac{\partial q}{\partial x} = i_e \qquad (2.15)$$

and a friction equation of the form $q = \alpha y^m$ $\qquad (2.16)$

where m is a coefficient and α is a function of the water properties, surface roughness, bed slope and gravity. Equations (2.15) and (2.16) apply to a wide flat bottom channel or overland flow. The flow q is per unit width and flow depth is y.

The quasi-steady flow approximation was originally termed the kinematic wave approximation since waves can only travel downstream and are represented entirely by the continuity equation. Since the dynamic forces are omitted, the Froude number $F = v/\sqrt{(gy)}$ is irrelevant, and in fact that the wave speed c is not given by $C = \sqrt{gy}$ but may be derived by finding dx/dt for which $dy/dt = i_e$

where $\frac{dy}{dt} = \frac{\partial y}{\partial t} + \frac{\partial y}{\partial x}\frac{dx}{dt}$ $\qquad (2.17)$

From the friction equation (2.16) $\frac{\partial q}{\partial x} = \frac{\partial q}{\partial y}\frac{dy}{dx} = m\alpha y^{m-1}\frac{\partial y}{\partial x}$ $\qquad (2.18)$

Substituting into the continuity equation yields

$$\frac{\partial y}{\partial t} + m\alpha y^{m-1}\frac{\partial y}{\partial x} = i_e \qquad (2.19)$$

but since $i_e = dy/dt$, the left hand side of this equation must also equal dy/dt.

Therefore $\dfrac{dx}{dt} = c = m\alpha y^{m-1}$ (2.20)

which is the speed at which a wave of unvarying amplitude (if $i_e = 0$)
travels down the plane.

Since $v = \alpha y^{m-1}$, it may be deduced that the wave speed is related to
water velocity v by the equation; $c = mv$. (2.21)

KINEMATIC FLOW OVER IMPERMEABLE PLANES

The kinematic wave equations have an important advantage over the
dynamic and diffusion wave equations; analytic solutions are possible for
simple watershed geometries. In this section, the kinematic solutions are
developed for runoff from an impermeable rectangular plane. Under these
conditions, we are not concerned with estimating rainfall loss due to
infiltration, nor with routing flows first overland and then through a
complex stormwater drainage system. Numerical models generally are
required when infiltration is important or multiple routings are involved.

Rising Hydrograph-General Solution

For the case of a long impermeable plane, $A = by$, $Q = bq$ and $R = y$, where q is the flow per unit width, hence Eqs. 2.4 and 2.13 can be
written

$$\frac{\partial q}{\partial x} + \frac{\partial y}{\partial t} = i_e$$

(2.15)

and

$$q = \alpha y^m$$

(2.16)

where i_e is the rainfall excess intensity. Substituting Eq. 2.15 into Eq.
2.16 and performing the differentiation yields

$$\alpha m y^{m-1} \frac{\partial v}{\partial x} + \frac{\partial v}{\partial t} = i_e$$

(2.22)

Eq. 2.22 states that to an observer moving at the speed

$$\frac{dx}{dt} = \alpha m y^{m-1}$$

(2.23)

the depth of flow changes with the rainfall rate

$$\frac{dy}{dt} = i_e$$

(2.24)

Eqs. 2.23 and 2.24 provide the basis for a method of characteristics
solution to surface runoff. For steady rainfall excess intensity, Eq. 2.24
can be integrated to obtain

$$y = y_o + i_e t$$

(2.25)

where y_o is the initial water depth when rainfall begins. Eq. 2.25 is the equation for depth along each characteristic as that characteristic moves from some initial position toward the downstream end of the plane. The position on the characteristic at any instant in time is determined with Eq 2.23. For an initially dry surface $y_o = 0$, hence $y = i_e t$. Substituting this relationship into Eq. 2.23 gives

$$\frac{dx}{dt} = \alpha m (i_e t)^{m-1} \tag{2.26}$$

which integrates to

$$x = x_o + \alpha i_e^{m-1} t^m \tag{2.27}$$

or more simply

$$x = x_o + \alpha y^{m-1} t \tag{2.28}$$

which specifies the downslope position of the depth y after time t. x_o is the point from which the forward characteristics emanate, i.e., the origin of the characteristics at $t = 0$, and is measured from the upslope end of the plane.

The discharge at any point along a characteristic is given by the relationship

$$q = \alpha (i_e t)^m \tag{2.29}$$

Two characteristic paths are shown in Figure 2.3. The first emanates at a point interior to the plane and travels the distance $L-x_o$ during the time t_o. The depth and discharge at each point (x,t) along this characteristic is determined from Eqs. 2.25, 2.28 and 2.29. The second characteristic begins at the upslope end of the plane and travels the length of the plane during the time t_c. In this case, the depth at the upstream end is zero, $y_o = 0$, for all t. Therefore, as long as the rainfall intensity remains constant, once this initial characteristic has reached the downstream end of the plane, the depth profile along the plane will remain constant regardless of how long the rainfall persists, i.e., an equilibrium depth profile will be established. The time required for this to happen is the concentration time t_c. At equilibrium no additional rainfall is being added to surface detention storage, and the rate of outflow equals the rainfall rate.

Recognizing that generally what is required is the runoff hydrograph at the end of the plane catchment, the concept of an equilibrium time and flowrate suggests a way to simplify the use of Eqs. 2.25, 2.28 and 2.29. In the following sections, solutions and examples are given for the time to equilibrium, equilibrium depth profile and rising outflow hydrograph.

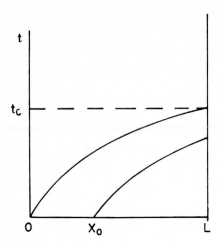

Fig. 2.3   Kinematic solution domain for plane catchment

Time of Concentration

One can solve for the time of concentration from Eq. 2.27 using the conditions that at $t = t_c$, $x - x_o = L$. Substituting and rearranging to solve for concentration time t_c which is equal to time to equilibrium t_m,

$$t_c = (L/\alpha i_e^{m-1})^{1/m} \tag{2.30}$$

For Manning-kinematic flow, time of concentration in minutes is

$$t_c = (6.9/i_e^{0.4})(nL/S_o^{0.5})^{0.6} \tag{2.31}$$

for i_e in mm/hr and L in metres and

$$t_c = (0.928/i_e^{0.4})(nL/S_o^{0.5})^{0.6} \tag{2.32}$$

for i_e in in/hr and L in feet

EXAMPLE 1. Estimate the time of concentration for a rainfall rate of 25 mm/hr on an asphalt parking lot 50 metres long and sloped at 1%. Assume n = 0.023.

Using Eq. 2.31, we find

$$t_c = \frac{6.90}{(25)^{0.4}} \left[\frac{0.023(50)}{(0.01)^{0.5}} \right]^{0.60} = 8.2 \text{ minutes}$$

Hence, a rain intensity of 25 mm/hr will bring the parking lot to equilibrium in 8.2 minutes.

Equilibrium Depth Profile

An expression for the equilibrium depth profile is found by solving Eqs. 2.25 and 2.28 simultaneously, and recalling that at $x_o = 0$, $y_o = 0$.

The resulting expression is

$$y(x) = (i_e x/\alpha)^{1/m} \tag{2.33}$$

which for Manning-kinematic flow in SI units becomes

$$y(x) = (ni_e x/S_o^{0.5})^{0.6} \tag{2.34}$$

EXAMPLE 2.   Estimate the equilibrium depth at the end of the asphalt parking lot in Example 1.

We need to be careful with units. The rainfall rate is in mm/hr; but the units implicit in the Manning equation are metres and seconds. Therefore, we need to divide the rainfall rate by 3.6×10^6.

$$i_e = 25/(3.6 \times 10^6) = 6.9 \times 10^{-6} \text{ m/sec}$$

From Eq. 2.34

$$y(L) = \left[\frac{0.023(6.9 \times 10^{-6})(50)}{(0.01)^{0.5}} \right]^{0.6} = 0.0034 \text{ metres}$$

or $y(L) = 3.4$ mm

The Receding Hydrograph

Henderson and Wooding (1964) derived the kinematic equations for the falling hydrograph. There are two cases involved: I. when the rising hydrograph is at equilibrium, and II, when the rising hydrograph is at a flow less than equilibrium, i.e., partial equilibrium.

Case I.   Duration of rainfall, $t_d \geq t_e$. After the rainfall stops, from 2.24, it can be seen that on a characteristic

$$dy/dt = 0 \tag{2.35}$$

which integrates to $y = c$, where c is some constant. Substituting this relationship into Eq. 2.23 reveals that the corresponding characteristic trajectories are lines parallel to the plane and that the depth, discharge and wave speed dx/dt, remain constant along a characteristic. This means that beginning with a point on the equilibrium profile and realizing that the future coordinates of that depth will lie on a single characteristic, Eq. 2.23 can be used to locate the point in space at any future time. This principle is illustrated in Figure 2.4.

The equilibrium depth profile at the cessation of rainfall is indicated as the line A-B_1-C_3. After some time Δt the depth profile is A-B_2-C_2. The depth at point B_1, y_1, has moved along a constant characteristic path to the point B_2.

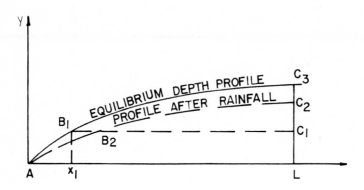

Fig. 2.4   Water depth profile

The distance moved is given by

$$\Delta x = \alpha m y^{m-1} \Delta t \tag{2.36}$$

The new x co-ordinate is

$$x = x_1 + \Delta x \tag{2.37}$$

$$= x_1 + \alpha m y_1^{m-1} (t-t_d) \tag{2.38}$$

where x_1 was the position for point B_1. Note that if the storm duration $t_1 > t_c$ the time to equilibrium, then $x_1 = x_e$. That is, once the equilibrium depth profile is established it will remain constant as long as the rainfall continues at a steady rate. From Eq. 2.33, the equilibrium depth can be expressed as

$$y_1 = \left[i_e x_1 / \alpha \right]^{1/m} \tag{2.39}$$

Substituting this relationship into Eq. 2.38 gives

$$x = x_1 + \alpha m \left[i_e x_1 / \alpha \right]^{(m-1)/m} (t-t_d) \tag{2.40}$$

At the downstream end of the plane $x = L$ and $q = \alpha y_1^m = i_e L$.
After substituting these identities into Eq. 2.40, we obtain the following relationship between discharge and time for the recession hydrograph

$$q - i_e L + i_e m \alpha^{1/m} (q)^{(m-1)/m} (t-t_d) = 0 \tag{2.41}$$

Case II. Duration of rainfall, $t_d < t_c$. If the rain stops prior to reaching equilibrium, then the depth profile at $t = t_d$ would correspond to one similar to $A-B_1-C_1$ in Figure 2.4. That is, an equilibrium depth profile will be developed from the upslope end of the plane at $x=0$ to some point x_1 given by

$$x_1 = \alpha y_1^{m-1} t_d \qquad (2.42)$$

The depth at point B_1 will move at a constant rate and will reach the end of the plane at time t_*. This time is evaluated as

$$t_* = t_d + \frac{L - x_1}{dx/dt} \qquad (2.43)$$

Incorporating Eqs. 2.26, 2.27, and 2.30, Eq. 2.43 becomes

$$t_* = t_d + \frac{\alpha i_e^{m-1} t_c^m - \alpha i_e^{m-1} t_d^m}{\alpha m i_e^{m-1} t_d^m} \qquad (2.44)$$

which can be simplified to

$$t_* = t_d \left\{ 1 + \frac{1}{m} \left[(t_c/t_d)^m - 1 \right] \right\} \qquad (2.45)$$

The discharge at the end of the plane will remain constant between $t_d \leq t \leq t_*$ and will be

$$q = \alpha (i_e t_d)^m \qquad (2.46)$$

After t_*, the recession proceeds according to Case I and Eq. 2.41 applies

EXAMPLE 3. Determine the runoff hydrograph from the parking lot in Example 1 for the same rainfall rate but of 10 minutes duration. Use the Manning kinematic solution.

The solution requires that we first determine the time to equilibrium which was done in Example 1. The next step is to generate the rising hydrograph. If $t_d \gtreqless t_c$ the rising hydrograph will be an equilibrium hydrograph. Finally, we must determine which case for recession applies and then determine the recession graph accordingly.

From Example 1 we know $t_c = 8.2$ minutes, therefore this event satisfies the conditions for an equilibrium rising hydrograph and Case 1 recession. Because $t_d \gtreqless t_c$, an equilibrium profile will exist on the plane during the time interval from $t=8.2$ minutes until $t=10$ minutes. During that time runoff from the plane will be constant and equal to the peak rate. The rising graph is given by Eq. 2.29 and the recession graph by Eq. 2.41.

First determine the equation for the rising graph. The coefficient in Eq. 2.29 is

$\alpha = S^{0.5}/n = (0.01)^{0.5}/0.023 = 4.35$

The depth in metres is determined by

$y = i_e t/(6 \times 10^4)$

where t is in minutes; and the discharge in $m^3/sec/m$–width of plane is

$q = 4.35 \, [(i_e t)/(6 \times 10^6)]^{5/3}$

Next, determine the equation for the recession hydrograph. After the appropriate substitutions and units conversion, Eq. 2.41 becomes

$$q - \frac{25(50)}{3.6 \times 10^6} + \frac{25}{3.6 \times 10^6}(5/3)\,(4.35)^{0.6}(q)^{0.4}(60)\,(t-10) = 0$$

TABLE 2.1   Runoff Hydrograph Ordinates

Time, Minutes	Depth, mm	Discharge, m^3/sec
0.0	0.0	0.0×10^5
1.0	0.42	1.0
2.0	0.83	3.2
3.0	1.25	6.3
4.0	1.67	10.2
5.0	2.08	14.8
6.0	2.50	20.0
7.0	2.92	25.9
8.0	3.33	32.4
9.0	3.42	33.7
10.0	3.42	33.7

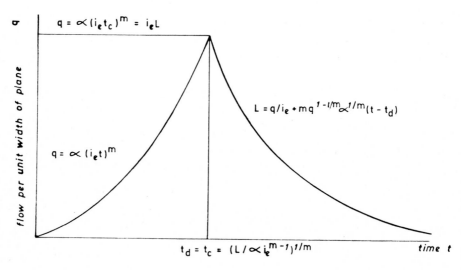

Fig. 2.5   Kinematic hydrograph shape for simple plane with $t_d = t_c$

FRICTION EQUATION

One of the kinematic equations is a friction energy loss equation. There are many friction equations in use in hydraulic engineering and a generalized comparison is made below. The most popular equation relating flow rate to friction energy gradient is perhaps that of Manning, which may be written as

$$Q = \frac{K_1}{n} AR^{2/3}S^{\frac{1}{2}}$$
(2.47)

where K_1 is 1 in S.I. units (metre-kilogram-seconds) and 1.486 in English units (foot-pounds-seconds), n is the Manning roughness, A is the cross sectional area, R the hydraulic radius A/P, P the wetted perimeter and S the energy slope. The S.I. system of units is adopted below but it should be noted that the equation is not dimensionless, and the roughness factor n is a function of gravity. Written in terms of flow per unit width of a wide rectangular channel, (as for an overland flow plane)

$$q = \frac{\sqrt{s}}{n} y^{5/3}$$
(2.48)

since hydraulic radius $R \stackrel{\sim}{=} yb/b = y$ and area $yb = y$. Hence $\alpha = s^{\frac{1}{2}}/n$ and m = 5/3. The Manning roughness coefficient n is reputedly a constant for any surface roughness. This holds for large Reynolds numbers and fully developed turbulent flow, but comparison with the Darcy Weisbach equation indicates that n actually increases for low Reynolds numbers ($yv/v < 1000$ where v is the kinematic viscosity of water, about $10^{-6} m^2/s$). The Manning equation may be compared with the Darcy equation employing Strickler's equation for roughness, $n = 0.13K_1 k^{1/6}/g^{\frac{1}{2}}$ where k is a linear measure of roughness analogous to the Nikuradse roughness for pipes (in metres if $K_1 = 1$). Substituting into Manning's equation yields

$$Q = 7.7(R/k)^{1/6}A(SRg)^{\frac{1}{2}}$$
(2.49)

If this equation is compared with Darcy's equation in the form

$$Q = (8/f)^{\frac{1}{2}} A(SRg)^{\frac{1}{2}}$$
(2.50)

it will be seen Strickler in effect used a Darcy friction factor equal to $0.135(k/R)^{1/3}$. (Note British practice is to use λ in place of f as they use f for a different factor.) According to Colebrook and White,

$$\frac{1}{\sqrt{f}} = -2 \log \left(\frac{k}{14.8R} + \frac{2.5}{Re \ f^{\frac{1}{2}}} \right)$$
(2.51)

where Re is the Reynolds number, for pipes VD/ν, or $4VR/\nu$ for non circular cross sections. Whereas the Colebrook-White equation predicts higher values of the Darcy friction coefficient f for low Reynolds number and any relative roughness k/R, the Strickler equation assumes f depends only on the relative roughness k/R. The Strickler and Manning equations can therefore be expected to underpredict roughness for low Reynolds numbers. Higher values of n should therefore be used for overland flow than for channel flows.

In general, the value of n and hence flow depth has to be determined by trial (assuming the Colebrook-White equation to apply and not Strickler's). It is therefore probably easier to use the Darcy equation for this purpose but since an explicit equation is required for analytical solutions to the kinematic equations and the variation in n is less than the variation in f with y, the Manning equation is preferred.

Table 2.2 indicates values of n and f with varying water depths in a wide channel with a slope of 0.0025 and absolute roughness k = 0.0125 m. The values of f are calculated from the Colebrook-White equation using a first estimate of Re from Manning's equation, and then n is re-calculated from $n = (f/8g)^{\frac{1}{2}}R^{1/6}$ i.e. as for Strickler's equation.

TABLE 2.2 - Variation of f and n with depth

Water Depth, m	Reynolds No.	Darcy f	Manning n
1.0	2×10^6	0.03	0.02
0.1	50 000	0.09	0.023
0.01	1 000	0.60	0.04

The Chezy equation is often used in preference to the Manning equation in American practice. This equation is

$$v = K_2 C\sqrt{(RS)} \qquad (2.52)$$

where C is known as the Chezy coefficient and K_2 is 1 in ft - second units and 0.552 in S.I. units. In fact the Chezy equation is very similar to the Darcy equation in the form

$$V = \sqrt{(8g/f)} \sqrt{(RS)} \qquad (2.53)$$

and it will be seen that $C = \sqrt{(8g/f)}/K_2$,

also for turbulent flow from (2.51) $1/\sqrt{f} \cong 2 \log (14.8R/k)$ \qquad (2.54)

Hence $C = (2/K_2) \sqrt{8g} \log (14.8/k)$ (2.55)

or $v \cong \sqrt{32gRS} \log (14.8/k)$ (2.56)

This equation stems from the log velocity distribution across a section whereas the Strickler equation follows from a 1/6 power law fit to the velocity distribution.

Resistance to rainfall induced overland flow over natural and man-made surfaces is influenced by several factors including surface roughness, raindrop impact, vegetation, wind and infiltration. Although there have been many laboratory and field investigations to determine the relative importance of these factors, the appropriate resistance formula, and methods for parameter estimation, in practice the convention has been to use either the Darcy-Weisbach equation modified for raindrop impact, or the traditional forms of the Manning or Chezy equations.

In laminar flow studies of overland flow the approach has been to assume the Darcy-Weisbach resistance law is appropriate, i.e.

$$v = \sqrt{(\frac{8g}{f} Sy)}$$

and to estimate the friction factor, f, from the theoretical relationship between f and Reynolds number Re,

$f = K/Re$ (2.57)

where K is a parameter related to the surface roughness characteristics and raindrop impact. The parameter K is approximated by

$K = K_o + Ai^b$ (2.58)

where K_o is the parameter for surface roughness and A and b are empirical parameters. When i is in inches per hour, the coefficient A is of the order of 10 and the exponent b is approximately unity. Typical values for K_o, Manning's n, and Chezy's C are given in Table 2.3. These values are ranges found in the literature and were obtained utilizing data from controlled experiments or from small experimental watersheds.

TABLE 2.3   Overland Flow Resistance Parameters

Surface	Laminar Flow K_o	Turbulent Flow Manning n	Turbulent Flow Chezy C
Concrete or Asphalt	24 – 108	0.01 – 0.013	73–38
Bare Sand	30 – 120	0.01 – 0.016	65–33
Gravelled Surface	90 – 400	0.012 – 0.03	38–18
Bare Clay to Loam Soil	100 – 500	0.012 – 0.03	36–16
Sparse Vegetation	1000 – 4000	0.053 – 0.13	11–5
Short Grass	3000 – 10000	0.10 – 0.20	6.5–3.6
Bluegrass Sod	7000 – 40000	0.17 – 0.48	4.2–1.8

In the case of turbulent flow, either the Manning or Chezy equation is used. The Manning equation is probably the more popular equation and is used more often in watershed simulation studies. The reasons for this are obviously its wide-spread acceptance in open channel hydraulics and the availability of extensive tables of n-values for most channel types and conditions.

Reported research indicates that low flows are laminar and that high flows are turbulent; but the location of the transitional Reynolds number is indeterminate which makes it difficult to apply the Darcy-Weisbach resistance formulation throughout the entire hydrograph. Transition from laminar to turbulent flow has been reported at Reynolds numbers ranging from 20 to 2,000; with the range $300 < Re < 500$ being the most frequently reported.

Overton (1972) analyzed 214 equilibrium hydrographs from an earlier study of airfield drainage conducted by the U.S. Army Corps of Engineers (1954). He noted that these hydrographs supported the argument for low flows being laminar and high flows being turbulent. In almost every case, the rising hydrograph initially rose very slowly indicating viscous laminar flow, and then became turbulent as the flow increased. Overton analyzed the rising portion of all 214 hydrographs in dimensionless form. He normalized the discharge by the rain rate, and time by a lag time parameter, t_L, which he defined as the time from the occurrence of 50% of the rainfall to 50% of the runoff volume. The normalized (dimensionless) hydrographs were plotted on transparent paper and superimposed. It was apparent, that within a small error, a single dimensionless rising hydrograph could represent all 214 hydrographs. The average dimensionless rising hydrograph was then plotted against the laminar, Manning and Chezy dimensionless rising hydrographs as shown in Figure 2.6. Flows appear to be laminar during the first half of the period of rise and turbulent during the second half. An error analysis indicated a 15% standard error in fitting the entire rising hydrograph for the Manning kinematic solution, and 19% for both the laminar and Chezy solutions.

To illustrate the effect of using only the Manning equation to represent the flow equation hence the flow resistance, throughout the entire hydrograph, consider Izzard's (1946) laboratory experimental run Nos. 136 and 138. Run No. 136 consisted of two bursts with rainfall intensity of 3.56 in/hr interrupted by a two minute lull. The first burst lasted for 10 minutes and the second, 11 minutes. The rainfall event produced a maximum Froude number of 0.55 and a minimum kinematic flow number of 156. Run No. 138 consisted of two bursts: the first was 1.83 in/hr for 8 minutes and the second was 3.55 in/hr for 8 minutes. This

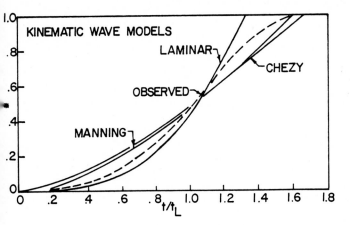

Fig. 2.6   Comparison of Turbulent and Laminar Kinematic Wave
            Solutions with Observed Rising Hydrographs.

event produced approximately the same maximum Froude number and minimum kinematic flow number as Run No. 136. The runoff surface was an asphalt plane with the following physical characteristics:

L = 72 ft; Manning n-value = 0.024; and S_o = 0.01.

Blandford and Meadows (1983) analyzed these events with a finite element formulation of the kinematic overland flow model and obtained the results shown in Figure 2.7. For Run No. 136, the predicted rising and falling limbs of the hydrograph lag the observed lightly, while there is near perfect agreement at the higher flows. From Run No. 138, only the simulated falling limb lags the observed; the rest of the hydrograph matches the observed very well.

REFERENCES

Beven, K., Dec. 1982. On subsurface stormflow, Predictions with simple kinematic theory for saturated and unsaturated flows. Water Resources Res. 18 (6) pp 1627-33.
Blandford, G.E. and Meadows, M.E. 1983, Finite Element Simulation of Kinematic Surface Runoff, Proceedings, Fifth International Symposium on Finite Elements in Water Resources, University of Vermont, Benning- ton, Vermont.
Dunne, T. 1978. Field studies of hillslope flow processes. Ch. 7, Hillslope Hydrology, Ed. Kirkby, M.J. John Wiley, N.Y.
Henderson, F.M. and Wooding, R.A. 1964 , Overland Flow and Ground- water from a Steady Rainfall of Finite Duration, Journal of Geo- physical Research, 69 (8), pp. 1531-1540.

42

Izzard, C.F., 1946. Hydraulics of Runoff from Developed Surfaces, Highway Research Board, Proceedings of the 26th Annual Meeting pp. 129-150.

Morris, E.M., and Woolhiser, D.A., April 1980. Unsteady one-dimensional flow over a plane: partial equilibrium and recession hydrographs. Water Resources Research, 16(2) pp 355-366.

Overton, D.E., 1972. A Variable Response Overland Flow Model, Ph.D. Dissertation. Dept. of Civil Engineering, Univ. of Maryland.

Overton, D.E. and Meadows, M.E., 1976 Stormwater Modelling, Academic Press, New York.

St. Venant, A.J.C. Barre de, 1848. Etudes Theoriques et Pratiques sur le Mouvement de Eaux Courantes. (Theoretical and Practical Studies of Stream Flow), Paris.

U.S. Army Corps of Engineers 1954. Data Report, Airfield Drainage Investigations, Los Angeles District, Office of the Chief of Engineers, Airfields Branch Engineering Division, Military Construction.

Woolhiser, D.A. 1981. Physicallly based models of watershed runoff, pp. 189-202 in Singh, V.P., (Ed.) Rainfall Runoff Relationships. Water Resources Publications, Colorado, 582 pp.

Woolhiser, D.A. and Liggett, J.A., 1967. Unsteady one-dimensional flow over a plane. The rising hydrograph. Water Resources Research, 3(3) pp 753-771.

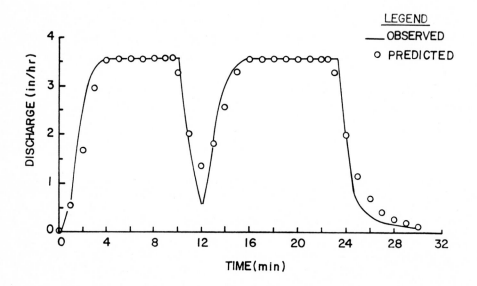

Fig. 2.7   Outflow Hydrograph for Izzard's Run No. 136

CHAPTER 3

HYDROGRAPH SHAPE AND PEAK FLOWS

DESIGN PARAMETERS

Knowledge of the runoff process enables flow rates and volumes to be predicted. Hydrograph characteristics are of interest to researchers, planners, designers and managers of drainage systems. The drainage engineer will be most concerned with peak flows for design purposes. It is also frequently useful to have the hydrograph shape especially if detention storage or routing can reduce the peak flow. Expressions for hydrograph shape and peak flows as a function of excess rainfall intensity can be derived as follows for overland flow off simple planes. The kinematic equations summarized below are used for this purpose.

Continuity

$$\frac{\partial q}{\partial x} + \frac{\partial y}{\partial t} = i_e \qquad (3.1)$$

Energy $\qquad q = \alpha y^m \qquad (3.2)$

where x is the direction of flow, t is time, i_e is the excess rainfall rate $i-f$, f is the loss rate and q is the discharge rate per unit catchment width. For the present all units must be assumed consistent. Later units will be introduced in order to render the numerical values more meaningful. It will be assumed in the following analysis that i and f are uniform in time and space for the duration of the storm t_d. q is the flow rate per unit width of plane, y is flow depth, α is a coefficient and m is an exponent.

SOLUTION OF KINEMATIC EQUATIONS FOR FLOW OFF A PLANE

The kinematic equations can be solved analytically for some simple cases. In particular the runoff from a rectangular plane catchment subject to uniform excess rain can be studied in detail and expressions for the time to equilibrium and hydrograph shape can be derived.

The following analysis demonstrates the simplicity of arriving at an equation for runoff for the catchment from a simple plane catchment sloping in the direction of flow. The analysis is handled more rigorously in chapter 2.

One starts with the generalized (one-dimensional) kinematic equations for overland flow namely 3.1 and 3.2.

If the Manning equation is assumed to hold then $\alpha = K_1 S_o^{\frac{1}{2}}/n$ and m = 5/3 where S_o is the slope of the plane in the direction of flow, and n is the Manning roughness. K is 1.0 in S.I. (metre) units and 1.486 in ft – sec units.

After rainfall commences, the water depth near the downstream end of the catchment will increase at a rate i_e, the excess rainfall rate. The water surface profile then will be parallel to the plane at the downstream end before equilibrium is reached, which is assumed to occur before the rain stops, i.e. $t < t_c \leq t_d$, where t_c is the time to equilibrium, usually referred to as the concentration time of the catchment.

Starting at the top or upstream end of the catchment where water depth and discharge rate will be zero, a negative surge due to a non-zero dy/dx will travel down the catchment overland increasing in depth as rain continues to fall. Then at any point in time downstream of the surge the water surface is increasing in depth at a rate i_e but upstream the water depth is at equilibrium since $\partial q/\partial x = i_e$ (see Fig. 2.4 line AB_1C_1).

Eventually the whole catchment will reach an equilibrium with input $i_e L$ per unit width equal to discharge q_L. At the instant the catchment reaches equilibrium

$$
\begin{aligned}
y_L &= i_e t_c \\
&= (q_L/\alpha)^{1/m} \quad \text{where } q_L = i_e L \\
\therefore t_c &= (L/\alpha i_e^{m-1})^{1/m}
\end{aligned}
\tag{3.3}
$$

t_c is the concentration time of the catchment, which is a function of the catchment length L, slope S_o, roughness n and excess rainfall rate i_e. The latter effect (i_e) rarely appears in time of concentration formulae associated with the rational method.

During the time of flow build-up the water depth at the exit is $i_e t$, and the corresponding discharge rate $q_L = \alpha(i_e t)^m$ $\hspace{1cm}$ (3.4)

The speed at which the reaction from upstream travels down the catchment before equilibrium, is obtained from the continuity equation. At the wave front the rate of increase in flow depth is

$$
\frac{dy}{dt} = \frac{\partial y}{\partial t} + \frac{\partial y}{\partial x}\frac{dx}{dt}
\tag{3.5}
$$

where dx/dt is the rate of travel of the wave front.

dy/dt $= i_e$ at the wave front point (and downstream of it). One also has from the continuity equation by expanding the $\partial q/\partial x$ term

$$
\frac{\partial y}{\partial t} + \frac{\partial q}{\partial y}\frac{dy}{dx} = i_e
\tag{3.6}
$$

By comparing with $\dfrac{\partial y}{\partial t} + \dfrac{dx}{dt}\dfrac{\partial y}{\partial x} = i_e$

one must have $\dfrac{dx}{dt} = \dfrac{\partial q}{\partial y}$ at the wave front and from 3.2, $dx/dt = m\alpha y^{m-1}$ (3.7)

Since $y = i_e t$,

$$\frac{dx}{dt} = m\alpha(i_e t)^{m-1} \tag{3.8}$$

which is the speed of the wave front at any time $t \leq t_c$. Also during equilibrium the discharge rate at any point x from the upstream watershed is $q = i_e x$ (line $AB_1 C_3$ in Fig. 2.4). Hence from 3.2

$$x = \alpha y^m / i_e \tag{3.9}$$

An expression for the discharge rate after the storm stops, which is assumed to be after the time to equilibrium ($t \geq t_d \geq t_c$), is obtained by considering the water depth profile along the catchment again. After the rain stops the effect of all upstream depths travels down to the exit at a speed dx/dt given by 3.7. To predict when the depth at the exit is 'y', imagine a series of waves travelling from the water profile curve in a downstream direction at a constant speed $dx/dt = \partial q/\partial y = m\alpha y^{m-1}$

Integrating, $x = x_o + m\alpha y^{m-1}(t-t_d)$ (3.10)

$$= q/i_e + mq^{1-1/m}\alpha^{1/m}(t-t_d) \tag{3.11}$$

since $y = (q/\alpha)^{1/m}$.   (3.12)

In particular at the exit,

$$L = q/i_e + mq^{1-1/m}\alpha^{1/m}(t-t_d) \tag{3.13}$$

which is an implicit expression for the falling limb of the hydrograph. The full hydrograph shape is thus as in Fig. 3.1.

HYDROGRAPHS FOR PLANES

Expressions for the rising and falling limbs of the hydrograph off a simple rectangular catchment were derived previously. The discharge at the mouth before time t_c or t_d is reached, is given by

$$q = \alpha(i_e t)^m \tag{3.14}$$

If rain continues after $t = t_c$ i.e. $t_d > t_c$, then the hydrograph top is horizontal as indicated in Fig. 3.1 case III.

If on the other hand rain stops at $t = t_c$ then the hydrograph falls immediately after t_c (case II). In either case it may be shown that the falling limb of the hydrograph is obtained from the implicit equation:

$$L = q/i_e + mq^{(1-1/m)}\alpha^{(1/m)}(t-t_d) \tag{3.13}$$

The total depth of excess rain has been kept constant in each case in Fig. 3.1 so that $i_e = p/t_d$ where p is the depth of precipitation.

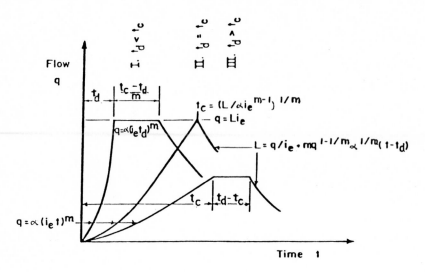

Fig. 3.1   Outflow hydrograph shape for different storm durations but similar total excess rain.

Also illustrated in Fig. 3.1 is the case of the hydrograph for a short storm $(t_d < t_c)$ (case I). After time t_d the downstream depth (and hence flowrate) remains constant until the influence of the upstream end reaches the exit.

The upstream limit of $y = i_e t_d$ is at

$$x = q/i_e = \alpha y^m/i_e = \alpha i_e^{m-1} t_d^m \qquad (3.15)$$

If this point travels a distance $L - x$ at a speed

$$dx/dt = m\alpha y^{m-1} \qquad (3.16)$$

it will reach the exit in time

$$\Delta t = \Delta x/m\alpha(i_e t_d)^{m-1} = (L - \alpha i_e^{m-1} t_d^m)/m\alpha i_e^{m-1} t_d^{m-1} = \frac{L/\alpha(i_e t_d)^{m-1} - t_d}{m} \qquad (3.17)$$

Assume i_{ec} is the excess rainfall rate for a storm duration equal to t_c. Since $i_e t_d = i_{ec} t_c$ for equal volume of rain, and

$$t_c = (L/\alpha i_{ec}^{m-1})^{1/m}$$

$$\Delta t = (t_c - t_d)/m \qquad (3.18)$$

This is the duration of the flat top of the hydrograph I in Fig. 3.1. It should be noted that the falling limbs of the hydrograph in Fig. 3.1 omit losses after rain stops. If infiltration (f) continues the hydrograph will look like those in Fig. 3.2. It is generally necessary to model such system numerically to get the hydrograph shape (see Wooding, 1965).

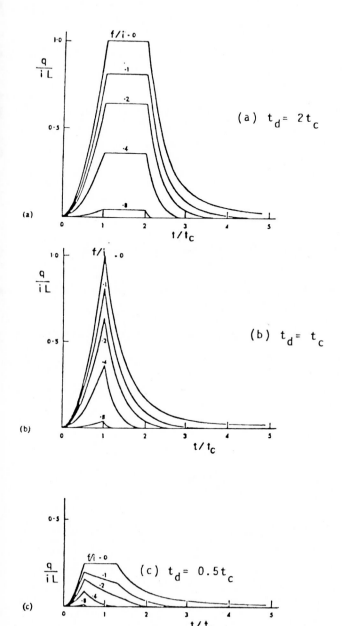

Fig. 3.2   Effects of infiltration on catchment discharge

DERIVATION OF PEAK FLOW CHARTS

If it can be assumed that the rainfall intensity-duration relationship for a specified frequency of exceedance can be approximated by a formula of the form

$$i = \frac{a}{(c+t_d)^p} \tag{3.19}$$

where i is the rainfall rate in mm/h or inches per hour and t_d is the storm duration in hours, then a simple estimate of peak flow can be derived. It is assumed that i is constant during the storm duration t_d and uniform over the catchment. The storm is also assumed stationary. c is a time constant unique for a particular rainfall region and p is an exponent, also a unique function of the region. Thus for temperate regions in South Africa it was found that $c = 0.24h$ and $p = 0.89$ while for coastal regions $c = 0.20h$ and $p = 0.75$ (Op ten Noort and Stephenson, 1982).

a is a function of rainfall region, mean annual precipitation, MAP, in mm and recurrence interval T in years e.g. $a = (b+e.MAP)T^{0.3}$ where b and e are regional constants. a is not dimensionless and if c and t_d are in hours, then i is in mm/h or inches per hour. An areal reduction factor is also necessary for large catchments (e.g. Stephenson, 1981).

Losses are subdivided into two components, an initial loss u in mm or inches and a uniform infiltration loss rate f in mm/h or inches per hour. A typical rainfall IDF (intensity-duration-frequency) relationship and the corresponding hyetograph with losses and excess runoff indicated is presented in Fig. 3.3.

The rate of excess rainfall is:

$$i_e = i - f \tag{3.20}$$

where f is the infiltration rate and the duration of excess rain is:

$$t_e = t_d - t_u = t_d - u/i \tag{3.21}$$

where u is the initial abstraction (measured in terms of a depth of rain).

For small catchments the maximum peak runoff rate occurs when the duration of excess rain equals the concentration time, t_c. For plain rectangular catchments the concentration time is a function of excess rainfall rate

$$t_c = (L/\alpha i_e^{m-1})^{1/m} \tag{3.22}$$

where $t_c = t_e$ (both in hours here) $\tag{3.23}$

$$\alpha = \sqrt{S}/n \text{ (S.I units) or } 1.486 \sqrt{S}/n \text{ (fps units)} \tag{3.24}$$

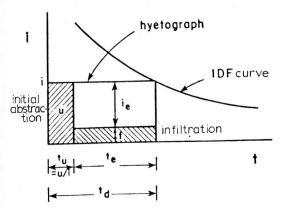

Fig. 3.3   Excess flow hyetograph derived from IDF curve

i_e is excess rainfall intensity, L is catchment length, S is the downstream slope, m is an exponent, 5/3 in Manning equation, n is the Manning roughness, q is the discharge rate per unit width.

The following expression may then be derived for i_e/a from equations (3.19) to (3.22):

$$i_e/a = \frac{1}{\left[c + \dfrac{u/a}{i_e/a+f/a} + \left(\dfrac{L}{\alpha a^{m-1}}\right)^{1/m} \left(\dfrac{a}{i_e}\right)^{1-1/m} \right]^p} - \frac{f}{a} \qquad (3.25)$$

Thus the maximum runoff rate per unit width of catchment

$$q = i_e L \qquad (3.26)$$

may be obtained in terms of a. Equation (3.25) may be solved iteratively e.g. using the Newton Raphson procedure, for i_e/a, or solution may be obtained with the aid of graphs of i_e/a plotted against $(L/\alpha\, a^{m-1})$. The storm duration t_d corresponding to the peak runoff may be obtained from (3.21) and (3.22).

Long Catchments

For very long catchments the theoretical concentration time t_c is high. In such cases the corresponding excess rainfall rate for a duration equal to t_c is low and in fact could conceivably be less than the infiltration rate f. It is thus apparent that in such cases the maximum runoff rate may coincide with a storm of shorter duration than the concentration time of the catchment. The entire catchment will thus not be contributing at the time of the peak in the (rising) hydrograph. If a local, intense storm turns out to be the design storm, the areal reduction factor applied to point rainfall intensity relationships may be less significant. The factor is generally closer to unity the smaller the lateral extent of the storm, but on the other hand shorter duration storms have a more significant reduction factor (less than long storms). These facts will not be revealed using the Rational method with rainfall proportional losses.

Before equilibrium is reached the runoff per unit width at the mouth of the catchment at any time t_e after the commencement of excess rain or runoff is

$$q = \alpha (i_e t_e)^m \tag{3.27}$$

where $t_e = t_d - u/i \tag{3.28}$

$$< t_c = (L/\alpha i_e^{m-1})^{1/m} \tag{3.29}$$

and $i_e = \dfrac{a}{(c+t_d)^p} - f \tag{3.30}$

hence $q = \alpha \{ (t_d - \dfrac{u}{i}) (\dfrac{a}{(c+t_d)^p} - f) \}^m \tag{3.31}$

or $\dfrac{q}{\alpha a^m} = \left[\{ t_d - \dfrac{u}{a} (c+t_d)^p \} \{ \dfrac{1}{(c+t_d)^p} - \dfrac{f}{a} \} \right]^m \tag{3.32}$

$q/\alpha a^m$ is plotted against t_d in Figs. 3.4 to 3.6 (the full lines) for different values of the dimensionless parameters $U = u/a$ and $F = f/a$. For all cases of $F > 0$ the lines exhibit a peak runoff and the corresponding storm duration t_d for an infinitely long catchment. For most catchments it is necessary to establish whether t_e is less than t_c i.e. whether the peak occurs before the catchment has reached equilibrium.

In fact, $t_e = t_d - t_u \tag{3.33}$

Therefore for $t_c = t_e$

$$t_d = t_u + t_c$$
$$= u/i + (L/\alpha)^{1/m}/i_e^{1-1/m} \tag{3.34}$$

or $\dfrac{L}{\alpha a^{m-1}} = \{ t_d - \dfrac{u}{a} (c+t_d)^p \}^m \{ \dfrac{1}{(c+t_d)^p} - \dfrac{f}{a} \}^{m-1} \tag{3.35}$

$L/\alpha a^{m-1}$ may therefore be plotted against t_d for selected values of u/a and f/a as on the right hand side of Figs. 3.4 to 3.6. Each chart is for a different initial abstraction factor U, and it may be necessary to interpolate between graphs for intermediate values of U. Lines for various F are plotted on each graph.

Now the peak runoff will be the maximum of either

(a)   that corresponding to $t_e = t_c$ for short catchments or

(b)   that for $t_e < t_c$ for long catchments.

In order to identify which condition applies, enter the chart for the correct $U = u/a$ with $L/\alpha a^{m-1}$ on the right hand side and using the dotted line corresponding to the correct F read off the corresponding $t_d = t_c + t_u$ on the abscissa. It may occur that the equilibrium t_c is off the chart to the right in which case it is probably of no interest since the following case applies. Select the full line with the $F = f/a$ and decide whether its maximum lies at or to the left of the value t_d previously established. If the peak lies to the left, read the revised design storm duration t_d corresponding to the peak, and the corresponding peak flow parameter $q/\alpha a^m$ on the left hand ordinate.

Modification for Practical Units

The preceding equations assume dimensional homogeneity. Unfortunately both the Manning resistance equations and the I–D–F relationships are empirical and the coefficients depend on the units employed. In the Manning form of equation (3.24), q is in m^2/s if $i_e t_e$ is in metres. α is $\sqrt{S}/n$ in S.I. units where S is the dimensionless slope and n is the Manning roughness.

It is most convenient to work with t_d in hours and i_e and a in mm/h. The numbers are then more realistic. In equation (3.32) if q is in $m^3/s/m$, α in m–s units, a in mm/h and t_d in hours then the left hand side should be replaced by

$$\frac{q1000^m}{\alpha a^m} = \frac{10^5 q}{\alpha a^{5/3}}$$

$$= \frac{10^5 Q}{B\alpha a^{5/3}} \tag{3.36a}$$

where Q is total runoff rate off a catchment of width B metres. Note that the right hand side of (3.32) is in h^m if t_d is in h, so no correction is made to the above factor to convert a to secs, only to convert mm to m. This is what the left hand axis of Figs. 3.4 to 3.6 represent if a is in mm/h. It is referred to as the runoff-factor, QF. Similarly the left hand side of equation (3.35) is $L/\alpha a^{m-1}$ in homogeneous units, or if a is in mm/h, L in metres and α in m–s units, then it should be replaced by:

$$\frac{L}{\alpha} \frac{a}{3600000} \frac{1000^m}{a^m} = \frac{L}{36\alpha a^{2/3}} \tag{3.37a}$$

This is termed the length factor LF.

If q is in $ft^3/s/ft$, α in ft–sec units, a in inches per hour and t_d in hours then the expression for Q should be replaced by

$$\frac{63q}{\alpha a^{5/3}} \tag{3.36b}$$

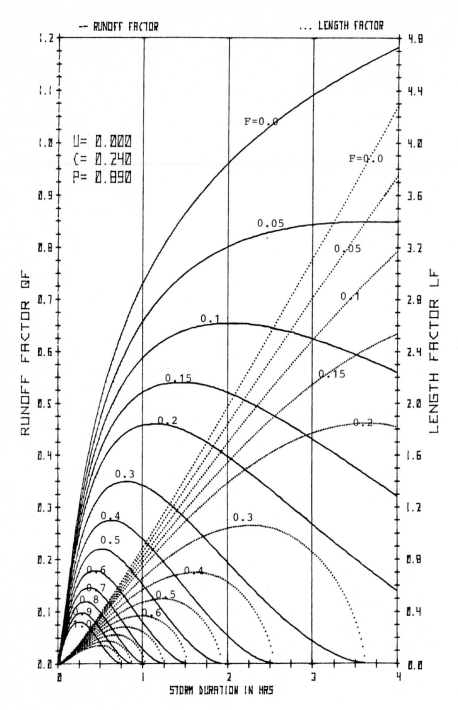

Fig. 3.4 Peak runoff factors for U = 0.00

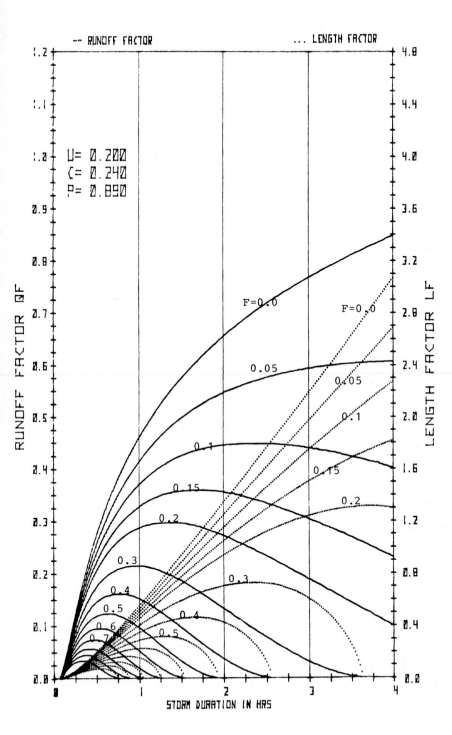

Fig. 3.5 Peak runoff factors for U = 0.20

54

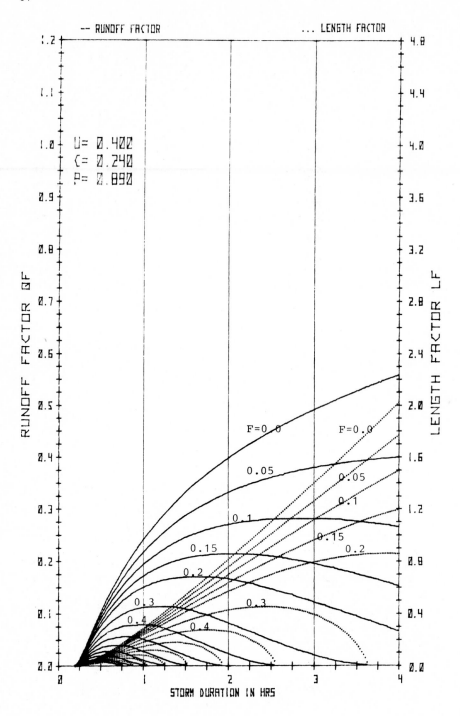

Fig. 3.6 Peak runoff factors for U = 0.40

and the length factor expression (3.37b) is replaced by

$$\frac{L}{687\alpha a^{2/3}}$$

Figs. 3.4 to 3.6 are plotted in terms of the dimensioned expressions for LF and QF, i.e. use t_d in hours and the other terms in the chosen metric or English units above. Further charts were published by Stephenson (1982).

EXAMPLE

Consider a plane rectangular catchment with the following characteristics:

overland flow length L = 800m

width B = 450 m

slope S = 0.01

Manning roughness n = 0.1

inland region, MAP = 620mm/annum

20 year recurrence interval storm

rainfall factor a = $(7.5 + 0.034 \times 620)20^{0.3}$ = 70mm/h

c = 0.24h, p = 0.89 in i mm/h = $a/(c+t_d)^p$

infiltration rate f = 14mm/h

initial abstraction u = 14mm

$\alpha = \sqrt{S}/n = 1.0$

F = f/a = 0.2

U = u/a = 0.2

$$LF = \frac{L}{36\alpha a^{2/3}} = \frac{800}{36 \times 1 \times 70^{2/3}} = 1.30$$

From Fig. 3.5 concentration time t_c = 3.0h

It will be noted that there may be two solutions for storm duration t_d. The longer one corresponds to a very low precipitation rate and is of little interest. Even the shorter time to equilibrium is longer than the storm resulting in the peak runoff. In this case it appears that the peak runoff corresponds to a 1.3 hour storm (shorter than the time to equilibrium) and the corresponding

QF = 0.30

   = $10^5 Q/B\alpha a^{5/3}$

therefore   Q = $0.30 \times 450 \times 1 \times 70^{5/3}/10^5$ = 1.60m³/s

The corresponding precipitation rate is:

i = $a/(c+t_d)^p$ = $70/(0.24 + 1.3)^{0.89}$ = 48mm/h

The equivalent rational coefficient C is

$1.60/(450 \times 800 \times 48/3600000)$ = 0.33

It may also be confirmed that the storm duration corresponding to time to equilibrium of the catchment is 2.8 hours:

If i_e = 70/(0.24+3.0)$^{0.89}$ -14 = 10.5mm/h = 2.9×10^{-6}m/s

t_c = $(L/\alpha i_e^{m-1})^{1-m}$

     = $\{800/1 \times (2.90 \times 10^{-6})^{2/3}\}^{3/5}$ = 9050s = 2.5h

t_u = u/i = 14/48   = 0.3h

Storm duration t_d = 2.8h

The corresponding runoff would be only:

i_eBL = 2.9 × 10^{-6} × 800 × 450 = 1.0m^3/s

i.e. the peak runoff corresponds to a storm of shorter duration than to time to equilibrium of the catchment.

EFFECT OF CANALIZATION

The charts presented are for the case of overland flow. It frequently occurs that runoff reaches channels, and thus flows to the mouth of the catchment faster than if overland. The critical storm duration may thus be shorter and the peak flow higher than with no canalization.

An estimate for the concentration time of a catchment with a wide rectangular channel down the middle may be made using this chapter if overland flow time can be neglected. The effective catchment width is taken as b, the stream width, and both rainfall rate i and losses f and u should be increased by the factor B/b where B is the true catchment width. The charts herein can then be applied as in the example.

In many situations both overland flow and stream flow are significant and the problem cannot be solved as simply as herein. The hydrologist must then resort to trial and error methods using dimensionless hydrographs for catchment – stream systems as presented later.

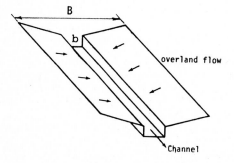

Fig. 3.7 Rectangular catchment with central collection channel.

ESTIMATION OF ABSTRACTIONS

The losses to be deducted from precipitation include interception on vegetation and roofs, evapotranspiration, depression storage and infiltration. The remaining losses may be divided into initial retention and a time-dependent infiltration. The losses are really functions of many variables, including antecedent moisture conditions and ground cover. Infiltration is time-dependent and an exponential decay curve is often used. The infiltration typically reduces from an initial rate of about 50 mm/h down to 10 mm/h over a period of about an hour. The rates, especially the terminal loss rate, will be higher for coarse sands than for clays.

The time-decaying loss rate could be approximated by an initial loss plus a uniform loss over the duration of the storm. Values of initial and uniform losses are tentatively suggested below. The mean uniform loss rates are average for storms of 30 minutes duration and the initial losses include the initial 10 minutes rapid infiltration or saturation amount. In the case of ploughed lands, and other especially absorptive surfaces an additional loss of up to 10mm or more may be included. Allowance must also be made for reduced losses from covered areas (paved or roofed).

TABLE 3.1   Surface Loss Parameters

	Initial abstraction					Infiltration rate	
	Surface retention			Max Soil moisture deficit			
		mm	inches	mm	inches	mm/h	inches/h
Paved	up to	5	0.2	0	0	0	0
Clay	up to	5	0.2	15	0.8	2 – 5	0.1 – 0.2
Loam	up to	7	0.3	20	1.2	5 – 15	0.2 – 0.6
Sandy	up to	10	0.4	30	1.5	15 – 25	0.6 – 1
Dense vegetation	up to	15	0.6	5	0.2	5 – 15	0.2 – 0.6

REFERENCES

Op ten Noort, T.H. and Stephenson, D., 1982. Flood Peak Calculation in South Africa. Water Systems Research Programme, University of the Witwatersrand, Report No. 2/1982.

Stephenson, D., 1981. Stormwater Hydrology and Drainage, Elsevier, Amsterdam. pp 276.

Stephenson, D., 1982. "Peak Flows from Small Catchments Using Kinematic Hydrology," Water Systems Research Programme, Report 4/1982. University of the Witwatersrand, Johannesburg.

Wooding, R.A., 1965. A Hydraulic Model for the Catchment-stream Problem, II, Numerical Solutions, Journal of Hydrology, 3, p 268-282.

CHAPTER 4

KINEMATIC ASSUMPTIONS

NATURE OF KINEMATIC EQUATIONS

The kinematic flow approximation has proved to be very useful in stormwater modelling and in the development of a better understanding of the runoff process. Kinematic models are deterministic models and represent a distributed, time-variant system. They can, therefore, be coupled with other process models to investigate the effects of land use change, temporal and spatial variations in rainfall and watershed conditions, and pollutant washoff.

Starting with the formulation of the kinematic wave theory by Lighthill and Whitham (1955), kinematic overland flow models have been utilized increasingly in hydrologic investigations. The first application to watershed modelling was by Henderson and Wooding (1964). The conditions under which the kinematic flow approximation holds for surface runoff were first investigated by Woolhiser and Liggett (1967); they found it is an accurate approximation to the full equations for most overland flow cases. Since then, analytical solutions have been obtained for runoff hydrographs during steady rainfall on simple geometric shaped watersheds; and numerical models have been developed for application to more complex watersheds and unsteady rainfall. With the easy availability of micro-computers, the numerical models are readily accessible. Successful use of these models requires a familiarity with computers and an understanding of kinematic overland flow.

KINEMATIC APPROXIMATION TO OVERLAND FLOW

Kinematic overland flow occurs when the dynamic terms in the momentum equation are negligible. There is no appreciable backwater effect and discharge can be expressed as a unique function of the depth of flow at all distance x and time t. That is,

$$Q = b\alpha y^m \tag{4.1}$$

where Q is the discharge, y is the depth of flow, b the width and α, m are constants.

The latter conclusion can be established by normalizing the momentum equation by the steady uniform discharge Q_n. The momentum equation then becomes

$$Q = Q_n \left[1 - \frac{1}{S_o} \left(\frac{\partial y}{\partial x} + \frac{v}{g} \frac{\partial v}{\partial x} + \frac{1}{g} \frac{\partial v}{\partial x} + \frac{q_i v}{gA} \right) \right]^{1/2} \qquad (4.2)$$

where S_o is the bed slope, v is flow velocity, g is gravity and A is cross sectional area. If the sum of the terms to the right of the minus sign is much less than one, then

$$Q \cong Q_n \qquad (4.3)$$

which means that gradually-varied flow may be approximated by a uniform flow formula such as Manning's equation. If one writes Manning's equation for a wide rectangular cross-section such as an overland flow plane, since the hydraulic radius can be approximated by the depth of flow, one obtains the following expression (SI units)

$$Q = \frac{1}{n} \, by \, y^{2/3} S_o^{\frac{1}{2}} \qquad (4.4)$$

or $\quad Q/b = \frac{1}{n} S_o^{\frac{1}{2}} y^{5/3} \qquad (4.5)$

where Q/b is discharge per unit width. For runoff from a plane surface with uniform roughness and slope, n and S_o are constant. Eq. 4.5 can therefore be written in the same form as Eq. 4.1 with $\alpha = S_o^{\frac{1}{2}}/n$ and $m = 5/3$; and for the cited conditions discharge can indeed be expressed as a unique function of the depth of flow.

Governing Equations

The governing equations for the kinematic overland flow approximation are Eq. 4.1 and the equation for continuity

$$\frac{\partial Q}{\partial x} + \frac{\partial A}{\partial t} = q_i \qquad (4.6)$$

where q_i is the inflow per unit length x.

Conditions for the Kinematic Approximation

The conditions under which the kinematic approximation holds for overland flow can best be illustrated by applying the full equations to runoff from a long, uniformly sloped plane of unit width as shown in Figure 4.1. The plane is of length L and slope S_o. Rainfall occurs over the plane at the rate $i(x,t)$, and infiltration is at the rate $f(x,t)$. By writing the rainfall and infiltration rates in terms of x and t, we include the effects of spatial and temporal variations in rainfall and soil. The continuity and momentum equations are written as

$$\frac{\partial y}{\partial t} + \frac{\partial (vy)}{\partial x} = i(x,t) - f(x,t) = i_e(x,t) \qquad (4.7)$$

and

In Terms of
x, & T

$$\frac{\partial v}{\partial t} + v\frac{\partial v}{\partial x} + g\frac{\partial y}{\partial x} = g(S_o - S_f) - \frac{vi_e}{y} \qquad (4.8)$$

where $i_e(x,t)$ is the rainfall excess rate at distance x and time t and the other terms have been defined previously.

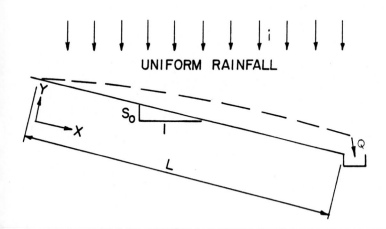

UNIFORM RAINFALL

Fig. 4.1 Uniform rain over a long impermeable plane

For the purpose of this discussion, S_f is conveniently defined by the Chezy equation

$$S_f = \frac{v^2}{C^2 y} \qquad (4.9)$$

C being the Chezy coefficient which equals $\sqrt{f/8g}$ where f is the Darcy friction factor. By writing Eqs. 4.7 and 4.8 in dimensionless form, the number of parameters are reduced from five to two with obvious advantages. Woolhiser and Liggett (1967) first presented the following dimensionless equations

$$\frac{\partial H}{\partial T} + U\frac{\partial H}{\partial X} + H\frac{\partial U}{\partial X} = 1 \qquad (4.10)$$

and

$$\frac{\partial U}{\partial T} + U\frac{\partial U}{\partial X} + \frac{1}{Fr_o^2}\frac{\partial H}{\partial X} = k(1 - \frac{U^2}{H}) - \frac{U}{H} \qquad (4.11)$$

where

$$H = y/y_o, \quad U = v/v_o, \quad X = x/L, \quad T = tv_o/L \qquad (4.12)$$

and y_o and v_o are the normal depth and velocity, respectively, at the end of the plane for a given steady rainfall excess rate, i_e. The normalizing parameters are related by:

$$i_e L = v_o y_o \qquad (4.13)$$

and

$$v_o^2/C_o^2 y_o = S_o \qquad (4.14)$$

The two independent parameters in Eqs. 4.10 and 4.11 are the normal flow Froude number, Fr_o, $v_o/\sqrt{(gy_o)}$, and the kinematic flow number, k (Woolhiser and Liggett, 1967).

$$k = \frac{S_o L}{y_o Fr_o^2} \qquad (4.15)$$

Woolhiser and Liggett (1967), Brutsaert (1968), Morris (1979) and Vieira (1983) solved Eqs. 4.10 and 4.11 for the rising hydrograph for a range of Fr_o and k values under normal depth, critical depth and zero depth gradient downstream boundary conditions. The solutions were started using an analytic solution for simple cases and numerical solutions in the other three characteristic solution zones. The results of all studies were quite similar. Sample results are shown in Figures 4.2 and 4.3.

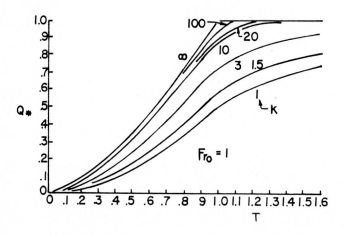

Fig. 4.2   Effect of varying k on dimensionless rising hydrograph (Woolhiser and Liggett, Water Resources Research, 3, 764, 1967, American Geophysical Union).

As seen in Figure 4.2, as the kinematic flow number increases, the solution converges very rapidly toward the solution for k equal to infinity. Woolhiser and Liggett noted that for $Fr_o = 1$ the maximum error between the rising hydrographs for k = 10 and k equal infinity is about 10 percent. The effect of varying Fr_o while holding k constant is shown in Figure 4.3. Similar to the results obtained for increasing k, as Fr_o increases, the solution converges to the solution for k equal to infinity.

What is the significance of k equal to infinity? If one divides Eq. 4.11 by k, the momentum equation reduces to the following expression.

$$1 - U^2/H = 0 \qquad (4.16)$$

Hence

$$U^2 = H \qquad (4.17)$$

Substituting Eq. 4.17 into the dimensionless continuity equation, Eq. 4.10, the kinematic wave equation is obtained.

$$\frac{\partial H}{\partial T} + \frac{\partial H}{\partial X}^{2/3} = 1 \qquad (4.18)$$

Solving Eq. 4.18 for an initially dry surface, we get

$$H = T \qquad (4.19)$$

and from Eq. 4.17

$$U = T^{1/2} \qquad (4.20)$$

Thus, the rising hydrograph is given by

$$Q_* = HU = T^{3/2} \qquad (4.21)$$

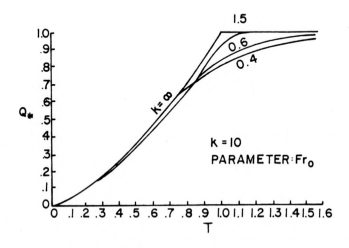

Fig. 4.3 Effect of varying Fr_o on dimensionless rising hydrograph (Woolhiser and Liggett, Water Resources Research, 3, 764, 1967, American Geophysical Union).

where Q_* is discharge normalized by the excess rainfall intensity. The result in Eq. 4.21 suggests that all rising hydrographs for steady excess rain on uniform planes can be represented by a single dimensionless hydrograph. This result also suggests there is a unique relationship between depth and discharge, and the depth is the normal depth for uniform flow at that discharge. When k is large the solution to the full equation can be closely approximated by the kinematic solution. This is the kinematic approximation which has been described in detail by several investigators (Lighthill and Whitham, 1955; Wooding, 1965; Woolhiser and Ligget, 1967; Morris and Woolhiser, 1980; Vieira, 1983).

Woolhiser and Liggett (1967) stated that the kinematic wave approximation may be used instead of the full equations if $k > 20$ and $Fr_o > 0.5$. Overton and Meadows (1976) noted that Eq. 4.21 is applicable to characteristic zone A (Fig. 5.1) and of the solution shown in Figure 5.1, zone A constitutes substantially all of the solution for kinematic flow numbers of 10 or greater. Therefore, they recommend the kinematic wave approximation be used only when $k > 10$, regardless of the Froude number value. Morris and Woolhiser (1980) re-evaluated Eqs. 4.10 and 4.11 and suggested that $Fr_o^2 k > 5$ if the kinematic approximation is used. It is interesting to note this is equivalent to the original criteria suggested by Woolhiser and Liggett, except that it allows the kinematic approximation to be used for $Fr_o < 0.5$, provided $k > 20$. With these conditions in mind, and using the results obtained in his own study, Vieira (1983) developed the plot in Figure 4.4 as a guide to determine when the kinematic and diffusion wave approximations may be used.

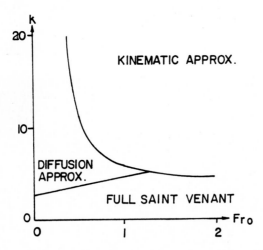

Fig. 4.4   Applicability of kinematic, diffusion and dynamic wave models (After Vieira, 1983)

Kinematic Flow Number

The kinematic flow number can be placed in terms of the physical and hydraulic characteristics of a plane by eliminating y_o and Fr_o from Eq. 4.15 using Eqs. 4.5 and 4.13. The resulting relationship is

$$k = \frac{gn^{1.2}S_o^{0.4}L^{0.2}}{i_e^{0.8}} \tag{4.22}$$

For rainfall intensity in mm/hr and length in meters, Eq. 4.22 becomes

$$k = 1.7 \times 10^6 \frac{n^{1.2}S_o^{0.4}L^{0.2}}{i_e^{0.8}} \tag{4.23}$$

and for rainfall intensity in in/hr and length in feet

$$k = 10^5 \frac{n^{1.2}S_o^{0.4}L^{0.2}}{i_e^{0.8}} \tag{4.24}$$

In general, high k values are produced on rough, steep, long planes with low rain rates.

Similarly, the quantity kFr_o^2 can be expressed in terms of the physical and hydraulic characteristics of a plane. From Eq. 4.15

$$kFr_o^2 = \frac{S_o L}{y_o} \tag{4.25}$$

If we write Eq. 4.21 in dimensional form using Eq. 4.12, Manning's resistance law instead of Chezy's, and the following definition for Q_*

$$Q_* = \frac{Q}{i_e} \tag{4.26}$$

we obtain the equation

$$Q = i_e A \left(\frac{t v_o}{L}\right)^{5/3} \tag{4.27}$$

where A is the contributing watershed area. For a steady rainfall excess rate, the flow is a maximum and equal to i_e when the terms inside the parentheses are equal to one, that is, when time is equal to the time of concentration, t_c, or preferably, the time to equilibrium. The quantity L/v_o is one definition for the time of concentration used in peak runoff estimates. According to Eq. 4.27, for a steady excess rate, at the time of concentration, the runoff rate is a maximum and equal to i_e. In other words, one definition for the time of concentration is that it is the time required for a watershed to reach equilibrium for a steady rainfall excess. This occurs when

$$t_c = L/v_o \tag{4.28}$$

Substituting Eq. 4.28 into Eq. 4.13 we get

$$y_o = i_e t_c \tag{4.29}$$

which, when substituted into Eq. 4.25 yields

$$kF_o^2 = \frac{S_o L}{i_e t_c} \tag{4.30}$$

Using the definitions of Eqs. 4.28 and 4.29, and Manning's equation, one obtains the desired expression, for rainfall in mm/hr,

$$kF_o^2 = \frac{1410 S_o^{1.3} L^{0.4}}{n^{0.6} i_e^{0.2}} \tag{4.30a}$$

and

$$kF_o^2 = \frac{460 S_o^{1.3} L^{0.4}}{n^{0.6} i_e^{0.2}} \tag{4.30b}$$

for rainfall in in/hr.

In general, kF_o^2 values are high for smooth, steep, long planes with low rainfall rates. This result is similar to the expression for k, except that the effect of roughness on the Froude number suggests the kinematic model may be more applicable to urban watersheds with smooth impervious surfaces.

To illustrate the hydrological applicability of these results, consider an asphalt parking lot with the following characteristics: L = 50 meters; S_o = 0.005; n = 0.022. For an average excess intensity of 50 mm/hr, k = 200 and kFr_o^2 = 31.

KINEMATIC AND NON-KINEMATIC WAVES

It was noted in Chapter 2 that the diffusion and kinematic wave models may be used instead of the full dynamic wave equations if certain assumptions can be made. In this section, conditions under which the two models can be applied to flood routing in streams are examined. The material presented here should give the reader a better under-standing of the physical nature of kinematic and non-kinematic waves.

The physical significance of kinematic and non-kinematic waves and the major differences between the respective models are better understood if the wave speed and crest subsidence (hydrograph dispersion) charac-teristics are known.

Wave Speed - Kinematic Waves

The kinematic wave speed is determined by comparing the continuity equation with no lateral inflow

$$\frac{\partial Q}{\partial x} + \frac{\partial A}{\partial t} = 0 \tag{4.31}$$

with the definition of the total derivative of Q

$$\frac{dQ}{dt} = \frac{\partial Q}{\partial x}\frac{dx}{dt} + \frac{\partial Q}{\partial t} \tag{4.32}$$

By rewriting Eq. 4.31 as

$$\frac{\partial Q}{\partial x} + \frac{\partial A}{\partial t}\frac{dQ}{dA}\frac{dt}{dx} = 0 \tag{4.33}$$

to an observer moving with wave speed, c,

$$c = \frac{dx}{dt} = \frac{dQ}{dA} \tag{4.34}$$

the flow rate would appear to be constant, i.e.,

$$\frac{dQ}{dt} = 0 \tag{4.35}$$

This result follows from the definition of the total derivative, Eq. 4.32, and the equation of continuity, Eq. 4.31.

For most channels where the flow is in-bank

$$\frac{dA}{dy} = B \tag{4.36}$$

where B is the channel top width in meters (feet); and since Q is a unique function of y

$$Q = \alpha y^m \tag{4.37}$$

the kinematic wave speed is given as

$$\frac{dx}{dt} = \frac{1}{B}\frac{dQ}{dy} = \frac{\alpha m}{B} y^{m-1} \tag{4.38}$$

This relationship is analogous to that of Seddon (1900) who observed that the main body of flood waves on the Mississippi River moved at a rate given by Eq. 4.38.

Eq. 4.38 implies that equal depths on both the leading and recession limbs of a hydrograph travel at the same speed. Since greater depths move at faster rates, it follows that the leading limb of the hydrograph will steepen and the recession limb will develop an elongated tail. Eq. 4.38 also shows that kinematic waves are propagated downstream only, i.e. Eq. 4.38 is a forward characteristic. Kinematic flow does not exist where there are backwater effects.

Crest Subsidence

Combining Eqs. 4.32 and 4.35, and substituting for Q using Eq.

4.37 it can be shown that to an observer moving with wave speed c

$$\frac{dy}{dt} = \frac{\partial y}{\partial x} \frac{dx}{dt} + \frac{\partial y}{\partial t} = 0 \qquad (4.39)$$

Manipulating this equation yields

$$\frac{dy}{dx} = \frac{\partial y}{\partial x} + \frac{\partial y}{\partial t} \frac{dt}{dx} = 0 \qquad (4.40)$$

which establishes that theoretically, the kinematic wave crest does not subside as the wave moves downstream.

These results show that a kinematic wave can alter in shape but does so without crest subsidence. Further, the maximum discharge rate occurs with the maximum depth of flow. (This is the assumption implicit in the slope–area method for estimating flood discharges from high water marks).

Hydraulic Geometry and Rating Curves

One important aspect of the kinematic wave model is the replacement of the momentum equation with a uniform flow formula, which is nothing more than a single valued rating between discharge and depth (or area) at a point in the stream. As discussed previously, the fact that natural channels are not prismatic leads to subsidence and dispersion of a hydrograph, suggesting that the discharge rating relationship is not unique but varies over the hydrograph. If the dispersive characteristics are small such that a variable rating relationship does not differ significantly from the single valued rating, the conclusion can be drawn that the main body of a hydrograph moves kinematically. In which case, the kinematic model (or the diffusion model) should be sufficient for most simulation purposes. This represents an economy of data and computational requirements over the dynamic wave model.

It is evident from the relationships between hydraulic geometry and discharge first set forth by Leopold and Maddock (1953) that the flow in many streams is essentially kinematic. The fact that the channel characteristics of natural streams seemed to constitute an interdependent system which could be described by a series of graphs having a simple geometric form suggested the term "hydraulic geometry". Subsequent studies have verified and expanded on this initial work with the result that hydraulic geometry equations may be used to estimate general channel characteristics at any location within the drainage system.

As a result of their analysis of the variation of hydraulic characteristics at a particular cross-section in a river, Leopold and Maddock proposed that discharge be related to other hydraulic factors in the following manner.

$$w = aQ^b \qquad (4.41a)$$

$$d = cQ^f \qquad (4.41b)$$

$$v = kQ^m \qquad (4.41c)$$

where w is width, d is depth, v is cross-sectional mean velocity, Q is discharge, and a, b, c, f, k, and m are best fit constants. It follows that since width, depth, and mean velocity are each functions of discharge, then $b + f + m = 1.0$; and $ack = 1.0$. Betson (1979) noted that a fourth relationship also can be presented

$$A = nQ^p \qquad (4.41d)$$

where A is the cross-sectional area of flow. Betson also noted that $f = p - b$ and $m = 1 - p$. The relationship in Eq. 4.41 are for individual stations in that they relate channel measures to concurrent discharge.

The results from several studies are shown in Table 4.1. It is notable that the values do not vary widely, particularly for the depth discharge relationship. These results reinforce the use of single valued rating curves and simplified routing models.

NON-KINEMATIC WAVES

The result in Eq. 4.40 frequently does not agree with nature. Rather, due to previously mentioned factors, flow peaks are seen to subside which suggests the application of the kinematic model is limited, and that either the diffusion or dynamic wave model is preferred. It is important then to examine the non-kinematic wave models and to establish how they differ from the kinematic model.

Differences between the two non-kinematic models can be investigated by examining the significance of each of the dynamic terms in the momentum equation. The discharge at a point in a stream is

$$Q = vA \qquad (4.42)$$

The momentum equation can be rewritten as follows:

$$\frac{Q}{A^2}\frac{\partial Q}{\partial x} - \frac{Q^2}{A^3}\frac{\partial A}{\partial x} + \frac{1}{A}\frac{\partial Q}{\partial t} - \frac{Q}{A^2}\frac{\partial A}{\partial t} + g\frac{\partial y}{\partial x} = g(S_o - S_f) - \frac{q_i}{A} \qquad (4.43)$$

The partial derivative of A with respect to time is removed in terms of the spatial derivative of Q using the continuity expression. After this substitution and rearranging, Eq. 4.43 becomes

$$\frac{2Q}{gA^2}\frac{\partial Q}{\partial x} - \frac{Q^2}{gA^3}\frac{\partial A}{\partial x} + \frac{1}{gA}\frac{\partial Q}{\partial t} + \frac{\partial y}{\partial x} = S_o - S_f \qquad (4.44)$$

TABLE 4.1   Typical Station Exponent Terms for Geomorphic Equations

Exponents

LOCATION OF BASIN	width b	depth f	velocity m	area p	Reference
Midwest	0.26	0.40	0.34	0.66	Leopold, et al. (1954)
Brandywine, P.A	0.04	0.41	0.55	0.45	ditto
158 Stations in U.S.	0.12	0.45	0.43	0.57	ditto
Big Sandy River, KY	0.23	0.41	0.36	0.64	Stall and Yang (1976)
Cumberland Plateau, KY	0.245	0.487	0.268	0.732	Betson (1979)
Johnson City, TN	0.08	0.43	0.49	0.51	Weeter and Meadows (1979)
Theoretical	0.23	0.42	0.35	0.67	Leopold and Langbein (1962)

At any cross-section Eq. 4.36 holds; and for most natural channels, the wave speed (celerity) is approximated by the kinematic wave speed. If Chezy's resistance equation is assumed

$$c = \frac{3Q}{2A} \tag{4.45}$$

Drawing on these two relationships and the definition for Froude number

$$Fr^2 = \frac{v^2}{gy} = \frac{Q^2 B}{gA^3} \tag{4.46}$$

the various terms in Eq. 4.44 can be rewritten as

$$\frac{2Q}{gA^2} \frac{\partial Q}{\partial x} = 3Fr^2 \frac{\partial y}{\partial x} \tag{4.47a}$$

$$- \frac{Q^2}{gA^3} \frac{\partial A}{\partial x} = - Fr^2 \frac{\partial y}{\partial x} \tag{4.47b}$$

and

$$\frac{1}{gA} \frac{\partial Q}{\partial t} = - \frac{9}{4} Fr^2 \frac{\partial y}{\partial x} \tag{4.47c}$$

Tracing back, the contribution of each term in the momentum equation is found (Meadows, 1981).

$$\frac{v}{g} \frac{\partial v}{\partial x} = (0.5 \, Fr^2) \frac{\partial y}{\partial x} \tag{4.48a}$$

and

$$\frac{1}{g}\frac{\partial v}{\partial t} = (-0.75 \ Fr^2) \ \frac{\partial y}{\partial x} \qquad\qquad (4.48b)$$

which allows the momentum equation to be written as

$$(1 - 0.25 \ Fr^2) \ \frac{\partial y}{\partial x} = S_o - S_f \qquad\qquad (4.49)$$

An equivalent expression was found by Dooge (1973).

Examination of Equations 4.48 and 4.49 reveals that the convective and temporal acceleration terms essentially are of equal magnitude but opposite sign, and hence, act to nearly cancel each other. These two terms are significant for Froude numbers greater than 0.60, where significance is taken as 10 percent of the coefficient value in Equation 4.49. Evidence of Froude numbers less than 0.60 for unsteady events in small streams is documented in the literature, e.g. (Gburek and Overton, 1973). Further, using the theoretical values for hydraulic elements of Leopold and Langbein (1962), it was shown by Meadows (1981) that

$$Fr \propto Q^{0.14}$$

demonstrating that Froude number is largely insensitive to increasing discharge in most natural streams for flow in bank. These results suggest the diffusion wave model can be confidently applied to most flood routing events.

Wave Speed

Based on the method of characteristics, it was shown that dynamic waves propagate both downstream (forward characteristic) and upstream (backward characteristic). The diffusion wave speed is given by the kinematic wave speed. As such, the diffusion wave model has only a forward characteristic meaning that wave forms are propagated only downstream and that backwater effects are negligible. It is left to the reader to confirm this.

Crest Subsidence

Both the dynamic and diffusive wave models simulate a dispersing hydrograph, hence, a subsiding wave crest. To illustrate, consider the modified diffusive wave equation, Eq. 4.49. For the following development, a rectangular cross section is assumed. As with the derivation of most overland and open channel flow equations, this assumption greatly simplifies the mathematics, yet does not alter appreciably the final form of the equations being developed.

Approximating the friction slope with Chezy's equation, Eq. 4.49 becomes

$$(1 - 0.25\ Fr^2)\ \frac{\partial y}{\partial x} = S_o - \frac{Q^2}{C^2 A^2 R} \tag{4.50}$$

Taking the partial derivative with respect to time

$$(1 - 0.25\ Fr^2)\ \frac{\partial}{\partial t}\ (\frac{\partial y}{\partial x}) = -\frac{Q^2}{C^2 A^2 R}\ [\ \frac{2}{Q}\ \frac{\partial Q}{\partial t} - \frac{2}{A}\ \frac{\partial A}{\partial t} - \frac{1}{R}\ \frac{\partial R}{\partial t}\ ] \tag{4.51}$$

From continuity

$$\frac{\partial Q}{\partial x} = q - \frac{\partial A}{\partial t} \tag{4.52}$$

or

$$\frac{\partial^2 Q}{\partial x^2} = \frac{\partial q}{\partial x} - \frac{\partial}{\partial x}\ (\frac{\partial A}{\partial t}) \tag{4.53}$$

Generally, over a reach, $\partial q/\partial x = 0$. Thus,

$$\frac{\partial^2 Q}{\partial x^2} = -\ \frac{\partial}{\partial x}\ (\frac{\partial A}{\partial t})\ = -\ \frac{\partial}{\partial t}\ (\frac{\partial A}{\partial x}) \tag{4.54}$$

For a prismatic section

$$\frac{dA}{dy}\ =\ B$$

such that

$$\frac{\partial A}{\partial x}\ \cdot\ B\frac{dy}{dA} \cong B\ \frac{\partial y}{\partial x} \tag{4.55}$$

which, when substituted into Eq. 4.54 yields

$$\frac{\partial^2 Q}{\partial x^2} = -\ B\ \frac{\partial}{\partial t}\ (\frac{\partial y}{\partial x}) \tag{4.56}$$

In obtaining Eq. 4.56, the assumption was made that $\partial B/\partial t = 0$; which is satisfactory if the channel is rectangular or the flood wave rises slowly. The momentum equation can now be written

$$(1 - 0.25\ Fr^2)\ \frac{1}{B}\ \frac{\partial^2 Q}{\partial x^2} = \frac{Q^2}{C^2 A^2 R}\ [\ \frac{2}{Q}\ \frac{\partial Q}{\partial t} - \frac{2}{A}\ \frac{\partial A}{\partial t} - \frac{1}{R}\ \frac{\partial R}{\partial t}\ ] \tag{4.57}$$

For a wide rectangular channel ($w > 10y$), the hydraulic radius, R, is approximately equal to the depth of flow, y. Using this approximation and continuity for a rectangular geometry

$$\frac{\partial Q}{\partial x} = q_i - w\ \frac{\partial y}{\partial t} \tag{4.58a}$$

or

$$\frac{\partial y}{\partial t}\ =\ \frac{q}{w} - \frac{1}{w}\ \frac{\partial Q}{\partial x} \tag{4.58b}$$

the right hand side of Eq. 4.57 is rewritten as

$$\frac{Q^2}{c^2 A^2 R} \left[\frac{2}{Q} \frac{\partial Q}{\partial t} + \frac{2}{A} \frac{\partial Q}{\partial x} + \frac{1}{A} \frac{\partial Q}{\partial x} - \frac{q_i}{A} \right]$$

Combining similar terms and recognizing that the coefficient terms are merely S_o,

$$S_o \left[\frac{2}{Q} \frac{\partial Q}{\partial t} + \frac{3}{A} \frac{\partial Q}{\partial x} - \frac{q_i}{A} \right]$$

The whole equation thus becomes

$$\frac{2}{Q} \frac{\partial Q}{\partial t} + \frac{3}{A} \frac{\partial Q}{\partial x} = \frac{(1 - 0.25 \, Fr^2)}{BS_o} \frac{\partial^2 Q}{\partial x^2} + \frac{3q_i}{A} \qquad (4.59)$$

Multiplying by $Q/2$

$$\frac{\partial Q}{\partial t} + \frac{3Q}{2A} \frac{\partial Q}{\partial x} = (1 - 0.25 \, Fr^2) \frac{Q}{2BS_o} \frac{\partial^2 Q}{\partial x^2} + \frac{Q3}{2A} q_i \qquad (4.60)$$

For Chezy's equation

$$c = \frac{3Q}{2A} \qquad (4.35)$$

Making this substitution into Eq. 4.60

$$\frac{\partial Q}{\partial t} + c \frac{\partial Q}{\partial x} = (1 - 0.25 \, Fr^2) \frac{Q}{2TS_o} \frac{\partial^2 Q}{\partial x^2} + cq_i \qquad (4.61)$$

which is a convective-diffusive equation for unsteady streamflow. This equation illustrates the origin of the diffusive wave label. The presence of the dispersion term (second partial derivative) confirms that the diffusive wave model simulates a subsiding peak.

One very interesting property of the crest region of a diffusive wave can be derived by rewriting Eq. 4.50 in terms of Q as follows

$$Q = CyB \sqrt{y[S_o - (1 - 0.25 \, Fr^2) \frac{\partial v}{\partial x}]} \qquad (4.62)$$

where hydraulic radius has been approximated by y. Taking the derivative with respect to x and equating to zero yields

$$3 \frac{\partial v}{\partial x} [S_o - (1 - 0.25 \, Fr^2) \frac{\partial y}{\partial x}] = y(1 - 0.25 \, Fr^2) \frac{\partial^2 y}{\partial x^2} \qquad (4.63)$$

In the region of the crest, the shape of the hydrograph is concave downward, and $\partial^2 y / \partial x^2 < 0$, and therefore, by Eq. 4.63, $\partial y / \partial x < 0$, also. That is, the peak flowrate does not occur where depth is a maximum, but at a point in advance of the maximum depth.

Looped Rating Curves

Eq. 4.62 clearly demonstrates that a single valued rating between discharge and depth (area) does not hold for non-kinematic waves. An approximate expression for the variable (looped) rating curve is given by

$$\frac{Q}{Q_n} = \sqrt{1 - \frac{(1 - 0.25\ Fr^2)}{S_o} \frac{\partial y}{\partial x}} \qquad (4.64)$$

where Q_n is the uniform flow at a given depth. This expression is rendered more useful if the spatial derivative is replaced by some alternate quantity, deductible from in-situ conditions.

Using the kinematic relationship

$$\frac{\partial y}{\partial x} = \frac{1}{c} \frac{\partial y}{\partial t} \qquad (4.65)$$

Eq. 4.64 can be written as

$$\frac{Q}{Q_n} = \sqrt{1 - \frac{(1 - 0.25\ Fr^2)}{S_o c} \frac{\partial y}{\partial t}} \qquad (4.66)$$

It must be noted that Eq. 4.66 is not strictly correct since the kinematic relationship was included.

A typical looped rating curve is shown in the Figure 4.5. Comparison with the associated discharge hydrograph illustrates that as a flood hydrograph passes a point, the maximum discharge is first observed, then the maximum depth, and finally a point where the flow is uniform. The uniform flow occurs when the flood wave is essentially horizontal and therefore has a slope, dy/dx, that is very small relative to the bed slope. This obviously will occur close to the region of maximum depth. The occurrence of uniform flow is illustrated graphically as the point of intersection of the looped rating curve with the single valued uniform flow rating curve.

It should be noted that the scale is exaggerated for clarity. The three points in question are more likely to occur much closer together than indicated by the figure.

The usefulness of the looped rating curve compared with a single valued rating curve is determined by how wide the loop is relative to the single valued curve. It should be noted however, that most published streamflow data and associated rating curves determined from field discharge measurements generally are better approximated by a single valued relationship. Looped curves can be approximated using Eq. 4.64 or 4.66 and time series records of river stage at a station.

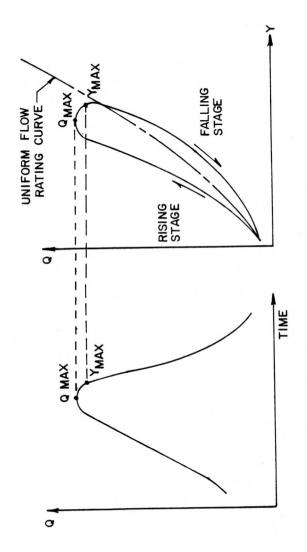

Fig. 4.5   Loop stage-discharge rating curve and associated discharge hydrograph for attenuating wave.

MUSKINGUM RIVER ROUTING

Flood routing refers to a set of models used to predict the temporal and spatial variations of a flood wave (runoff hydrograph) as it travels through a channel reach. Routing techniques are classed into two categories: hydraulic and hydrologic. The kinematic, diffusion and dynamic wave models are hydraulic routing models. The hydrologic models are based on continuity and an empirically derived relationship between channel storage and discharge; therefore, they are not as rigorous as the hydraulic models and represent a further simplification to the full equations for open channel flow.

Perhaps the best known and most widely used of the hydrologic models is the Muskingum routing model. This model was developed originally for flood routing on the Muskingum River in Central Ohio, hence the origin of the name. The model utilizes continuity

$$I + QL - O = \frac{dS}{dt} \tag{4.67}$$

where I is inflow to a river reach, QL is lateral inflow $(=q\Delta x)$, O is outflow and S is the storage within the reach; and the storage relationship

$$S = K\left[ZI + (1-Z)O\right] \tag{4.68}$$

where K is a characteristic storage time approximated as the travel time through a reach, and Z is a weighting coefficient.
For attenuating waves, $Z < 0.5$.

Equations 4.67 and 4.68 are solved using a finite differencing technique. Defining $I_1 = I(t)$ and $I_2 = I(t+\Delta t)$, and similarly, O_1, O_2, and S_1 and S_2, the following approximation to Eq. 4.67 is written

$$\frac{I_1 + I_2}{2} + QL - \frac{O_1 + O_2}{2} = \frac{S_2 - S_1}{\Delta t} \tag{4.69}$$

where QL is the average lateral inflow during the time interval Δt. The inflow hydrograph provides I_1 and I_2, and O_2 is the desired quantity. O_1 is known from either initial conditions or a previous calculation. S_1 and S_2 are expressed in terms of I and O as follows

$$S_2 - S_1 = K\left[Z(I_2 - I_1) + (1 - Z)(O_2 - O_1)\right] \tag{4.70}$$

Substituting Eq. 4.70 into Eq. 4.69 and simplifying gives

$$O_2 = C_0 I_2 + C_1 I_1 + C_2 O_1 + C_3 QL \tag{4.71}$$

where

$$C_o = \frac{-KZ + 0.5\Delta t}{K - KZ + 0.5\Delta t} \tag{4.72a}$$

$$C_1 = \frac{KZ + 0.5\Delta t}{K-KZ + 0.5\Delta t} \tag{4.72b}$$

$$C_2 = \frac{K-KZ - 0.5\Delta t}{K-KZ + 0.5\Delta t} \tag{4.72c}$$

and

$$C_3 = \frac{\Delta t}{K-KZ + 0.5\Delta t} \tag{4.72d}$$

K and t must have the same time unit, and the first three coefficients sum to 1.0.

Estimation of Model Parameters

The Muskingum model is quite sensitive to the selection of model parameters. Historically, K and Z have been estimated by matching model output with actual inflow-outflow records. The obvious shortcoming is that the model is limited to gauged streams. Oftentimes, we need to route flood hydrographs along ungauged streams. To do so requires that we have a means of estimating model parameters from available channel and hydrograph characteristics.

Following the technique of Cunge (1969) and using a Taylor series expansion to each of the terms in Eq. 4.67, it is transformed to an equivalent equation of the convective-diffusive form

$$\frac{\partial Q}{\partial t} + \frac{\Delta x}{K}\frac{\partial Q}{\partial x} = \left[\Delta x(1-Z)c(Q) - \frac{1}{2}\frac{\Delta x^2}{K}\right]\frac{\partial^2 Q}{\partial x^2} + \frac{QL}{K} \tag{4.73}$$

where Δx is the reach length. Comparison of this equation with Eq. 4.68 shows that

$$K = \frac{\Delta x}{c(Q)} \tag{4.74a}$$

and

$$Z = \frac{1}{2}\left[1 - \frac{Q(1-0.25Fr^2)}{BS_o\Delta xc(Q)}\right] \tag{4.74b}$$

Cunge (1969) and later researchers developed similar expressions to Eqs. 4.74. Ponce and Yevjevich (1978) considered the variation of K and Z with Q; while Dooge (1973) included the correction for dynamic effects, $(1-0.25Fr^2)$, in the equation for Z, but considered $c(Q)$ to be constant and not a function of Q. Therefore, Eqs. 4.74 are the most general expressions for K and Z (Meadows, 1981).

Another very important feature of Eq. 4.73 is that it demonstrates the Muskingum routing model is diffusive for $Z < 0.5$, and offers the same advantages of the diffusion wave model. If, however, $Z = 0.5$, the

Muskingum model predicts pure translation, and is equivalent to the kinematic wave model. Typical values for Z for natural streams are 0.3 to 0.4, and for prismatic channels, 0.4 to 0.5.

KINEMATIC AND DIFFUSION MODELS

We have discussed the kinematic and diffusion wave models as approximations to the dynamic wave model and have shown they are applicable for certain flood wave and channel conditions. As users, we need criteria or guidelines for selecting which model to use. Two notable works toward establishing such guidelines are Henderson (1963) and Ponce, et al. (1978).

Henderson conducted a theoretical examination of the governing equations similar to that presented in the previous sections. He compared theoretical results with a limited number of flood hydrographs, and noted that subsidence is most pronounced in the vicinity of the wave crest. Generally, he may be credited with efforts to classify flood waves according to the magnitude of S_o into waves broadly characteristic of steep, mild and intermediate slopes. However, he cautioned that this classification is not exhaustive, but should suffice for most floods in natural waterways. He did not offer specific guidelines to define mild, intermediate and steep, although he concluded the kinematic model is applicable in steep channels; the diffusion in mild and steep channels; and the dynamic to all three. The rationale for this conclusion is that he considered $Fr^2 >> 1$ in steeply sloped channels; hence, according to Eq. 4.49, the momentum equation will become $S_o = S_f$. For mild slopes, $Fr^2 << 1$, and the momentum equation becomes the strict diffusion wave model. For intermediate slopes all terms in the momentum equations are significant.

Ponce, et al., applied linear stability analysis in an effort to propose a theory that accounts for wave celerity as well as attenuation characteristics. To do so required they use a linearized, therefore somewhat simplified, version of the governing equations. Assuming a sinusoidal wave, they compared the propagation characteristics of the kinematic, diffusion and dynamic wave models. As expected, the dynamic wave model is applicable to the entire spectrum of waves that can be routed with a one-dimensional model. For Fr<2, the celerity of a dynamic wave is greater than the kinematic wave celerity. For Fr=2, roll waves will form. Thus, for primary waves (main body of a flood wave), Fr=2 is the threshold dividing attenuation and amplification. For secondary waves, Fr=1 is the threshold dividing the propagation upstream or downstream; for Fr=1 they remain stationary or propagate downstream only; and for Fr < 1, they propagate only downstream. A physical observation by Stoker (1957) explains this conclusion regarding secondary waves:

"What seems to happen is the following: small forerunners of a disturbance (wave) travel with the speed $\sqrt{gy}$ relative to the flowing stream, but the resistive forces act in such a way as to decrease the speed of the main portion of the disturbance far below the values given by $\sqrt{gy}$..."

Ponce, et al., did offer first generation criteria for application of the kinematic and diffusion wave models:

Kinematic:

$$T_B S_o \frac{v_o}{y_o} > 171$$

Diffusion:

$$T_B S_o (\frac{g}{y_o})^2 > 30$$

where T_B is the duration of the flood wave, S_o is the channel slope, v_p and y_o are the initial velocity and depth of flow, respectively, and g is gravity. Based on these criteria, the kinematic model applies to shallow flow on steep slopes (hence the steep channel of Henderson and surface runoff from hillslopes) and to long duration flood waves (slow rising floods on major rivers as observed by Seddon). The diffusion model is applicable to these as well as a wider range. When these two models break down, the dynamic model applies.

These criteria are significant in that they relate model application to channel slope and hydrograph characteristics. The reader is cautioned that these are only first generation formulae.

REFERENCES

Betson, R.P., 1979. A geomorphic model for use in streamflow routing, Water Resources Research, Vol. 15, No. 1, pp. 95–101.

Brutsaert, W. 1968. The initial phase of the rising hydrograph of turbulent free surface flow with unsteady lateral inflow. Water Resources Research, Vol. 4, pp 1189–1192.

Cunge, J.A., 1969. On the subject of a flood propagation computation method (Muskingum Method). J. Hydr. Res., Vol. 7, No. 2, pp. 205–230.

Dooge, J.C.I. 1973. Linear theory of hydrologic systems. U.S. Dept. of Agriculture, Agri. Res. Ser. Tech. Bull. No. 1968.

Gburek, W.J. and Overton, D.E., 1973. Subcritical kinematic flow in a stable stream. J. Hydr. Div. ASCE. Vol. 99, No. HY9, pp. 1433–1447.

Henderson, F.M., 1963. Flood waves in prismatic channels. J. Hydr. Div. ASCE, Vol. 89, No. HY4, pp. 39–67.

Henderson, F.M. and Wooding, R.A., 1964. Overland flow and groundwater from a steady rainfall of finite duration. Journal of Geophysical Research, Vol. 69, No. 8, pp. 1531–1540.

Leopold, L.B., et al., 1954. Fluvial Processes in Geomorphology. N.H. Freeman, San Francisco, Cal.

Leopold, L.B. and Langbein, W.B., 1962. The concept of entropy in landscape evolution. U.S. Geological Survey Prof. Paper 500-A.

Leopold, L.B. and Maddock, T., Jr. 1953. The hydraulic geometry of stream channels and some physiographic implications. U.S. Geological Survey Prof. Paper 252.

Lighthill, M.J. and Whitham, G.B., 1955. On kinematic waves: I. Flood movement in long rivers. Proc. Royal Society, London, Vol. 229, No. 1178, pp. 281 –

Meadows, M.E., 1981. Modelling the impact of stormwater runoff, in Proceedings, International Symposium on Urban Hydrology, Hydraulics and Sediment Control, University of Kentucky, Lexington, Kentucky, pp. 313-319.

Morris, E.M., 1979. The effect of the small slope approximation and lower boundary conditions on solutions of the Saint Venant Equation. Journal of Hydrology, Vol. 40, pp. 31-47.

Morris, E.M. and Woolhiser, D.A., 1980. Unsteady one-dimensional flow over a plane: partial equilibrium and recession hydrographs. Water Resources Research, 16 (2), pp 355-366.

Overton, D.E. and Meadows, M.E., 1976. Stormwater Modelling. Academic Press, N.Y.

Ponce, V.M., Li, R.M. and Simons, D.B., 1978. Applicability of kinematic and diffusion wave models. J. Hydr. Div. ASCE, Vol. 104, No. HY3, pp. 353-360.

Ponce, V.M. and Yevjevich, V. 1978. Muskingum-Cunge Method with variable parameters. J. Hydr. Div. ASCE, Vol. 104, No. HY12, pp. 1663-1667.

Seddon, J.A., 1900. River hydraulics. Trans: ASCE, Vol. 43, p. 179.

Stall, J.B. and Yang, C.T., 1970. Hydraulic geometry of Illinois streams. Research Report No. 15, Water Resources Research Center, University of Illinois, Urbana, Illinois.

Vieira, J.H.D., 1983. Conditions governing the use of approximations for the Saint Venant equations for shallow surface water flow. Journal of Hydrology, Vol. 60, pp. 43-58.

Weeter, D.W. and Meadows, M.E., 1978. Water Quality Modeling for Rural Streams. First Tennessee – Virginia Development District, Johnson City, 104 p.

Wooding, R.A., 1965. A hydraulic model for the catchment steam problem, I. Kinematic wave theory. Journal of Hydrology, Vol. 3, pp 254-267.

Woolhiser, D.A. and Liggett, J.A. 1967. Unsteady one dimensional flow over a plane – the rising hydrograph. Water Resources Research, Vol. 3, No. 3, pp. 753-771.

CHAPTER 5

NUMERICAL SOLUTIONS

METHODS OF SOLUTION OF EQUATIONS OF MOTION

There are no known general analytical solutions to the hydraulic equations

$$\frac{\partial Q}{\partial x} + \frac{\partial A}{\partial t} = q_i \qquad (5.1)$$

$$g\frac{\partial y}{\partial x} + v\frac{\partial v}{\partial x} + \frac{\partial v}{\partial t} = g(S_o - S_f) - q_i/A \qquad (5.2)$$

They must therefore be solved using the method of characteristics or numerical integration techniques. Available numerical methods include finite differencing and finite elements.

Finite differencing techniques are founded on the classical definition for a continuous derivative term. Use of these methods transforms the set of partial differential equations into an equal number of approximate algebraic equations which then are solved according to the rules of linear algebra.

The finite element method is a relatively recent approach to solving partial differential equations that govern hydraulic processes. The basis of finite element integration is approximating polynomials. In essence, the polynomial coefficients are adjusted to minimize an error term while satisfying known boundary conditions. The resulting polynomials express the unknown variables in terms of the known (independent) variables. The details of this method are beyond the scope of these notes; however, application of this method to kinematic overland flow is illustrated in a later chapter.

Given the many and varied ways of integrating the flood routing equations, one can logically ask which technique to choose. Some can be discarded as being inaccurate or unstable or too time consuming; others seem to reproduce solutions relatively well. However, there is no single answer to which method is "best". Indeed, the answer to that question depends on the particular application, and perhaps on the available computing equipment.

METHOD OF CHARACTERISTICS

The method of characteristics may be described as a technique whereby the problem of solving two simultaneous partial differential equations (continuity and momentum) can be replaced by the problem of solving four ordinary differential equations. This method has been known for many years; it was devised long before the computer as a means for graphically integrating the unsteady streamflow equations. The characteristic equations are no longer solved graphically, but are solved numerically using the computer.

By making the substitution

$$c^2 = gy \qquad (5.3)$$

into Eqs. 5.1 and 5.2 and then by writing first the sum, and then the difference, of the two new equations, we obtain the two equations

$$(v+c)\frac{\partial(v+2c)}{\partial x} + \frac{\partial(v+2c)}{\partial t} = g(S_o - S_f) - (v-c)\frac{q_i}{A} \qquad (5.4a)$$

$$(v-c)\frac{\partial(v-2c)}{\partial x} + \frac{\partial(v-2c)}{\partial t} = g(S_o - S_f) - (v+c)\frac{q_i}{A} \qquad (5.4b)$$

which are two equations in the form of directional derivatives of $v \pm 2c$. Recalling the definition of a total derivative, it can be shown that for

$$\frac{dx}{dt} = v \pm c \qquad (5.5)$$

then

$$\frac{dx}{dt}\frac{\partial(v\pm2c)}{\partial x} + \frac{\partial(v\pm2c)}{\partial t} = \frac{d(v\pm2c)}{dt} \qquad (5.6)$$

which gives the desired set of ordinary differential equations to replace the partial differential equations. The characteristic roots (directions) are given by Eqs. 5.5, and along each direction the respective total derivatives in Eqs. 5.6 hold. The resulting equations can be rewritten:

$$c^+ : \frac{dx}{dt} = v + c \qquad (5.5a)$$

$$\frac{d(v+2c)}{dt} = g(S_o - S_f) - (v - c)\frac{q_i}{A} \qquad (5.6a)$$

$$c^- : \frac{dx}{dt} = v - c \qquad (5.5b)$$

$$\frac{d(v-2c)}{dt} = g(S_o - S_f) - (v + c)\frac{q_i}{A} \qquad (5.6b)$$

where c^+ and c^- symbolically designate forward and backward characteristic respectively.

Physically, the characteristic roots represent the path in time and space followed by a disturbance, e.g. flood wave. The speed of propagation is given by the slope dx/dt; and the state of the system (values of the dependent variables) is given by the total derivatives that hold along the characteristic paths.

Mathematically, the characteristics are loci of possible discontinuities in the temporal and spatial derivatives of the dependent variables. Thus, one may think of a characteristic curve as a line of separation between two regions of somewhat different physical conditions. This is important when modelling unsteady flows where the boundary conditions vary with time, since the solution at interior points of the (x,t) domain are dependent on the boundary information. The solution at those points above a characteristic curve requires more boundary information than the solution below the characteristic. In fact, the characteristics in Eqs. 5.5 define four unique solution zones as shown in Figure 5.1.

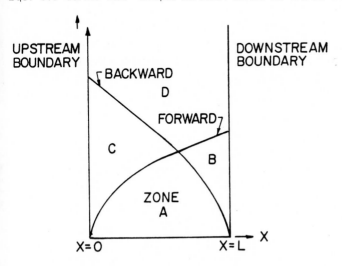

Fig. 5.1   Zones of solution domain defined by characteristics.
(Woolhiser and Liggett, Water Resources Research,
3, 755, 1967,   American Geophysical Union).

These zones are formed by the intersection of the forward and backward characteristics emanating from the upstream $(x=0)$ and downstream $(x=L)$ end of the channel reach at the initial time. The solution in Zone A requires only the initial values (the beginning state of the system at all x); while the solution in Zones B and C requires both initial values and a boundary condition. This is because these zones lie above the backward and forward characteristics, respectively. Zone B requires the downstream boundary condition, and Zone C, the upstream. Finally, Zone D, which lies above both characteristics requires the initial values and both boundary conditions.

Numerical Integration of Characteristic Equations

The objective of the method of characteristics is to fill the (x,t) plane with characteristics as shown in Figure 5.2. The unknowns are determined at the intersections where the four equations 5.5a, 5.5b, 5.6a and 5.6b are satisfied. The extent to which solutions can be obtained over the (x,t) plane is dependent on the amount of initial value, $(x,0)$, and boundary condition, $(0,t)$ and (L,t), information that is specified beforehand. Initial conditions are the velocity and depth of flow at all x at the beginning of the simulation, usually designated time zero. The upstream boundary condition typically is the known inflow hydrograph that is to be routed downstream. Values for v and y are obtained with known rating relationships. Usually the outflow hydrograph at the down-stream end (boundary) is the desired result; hence, Q, v and y at the downstream boundary are unknown. However, if rating relationships are known, they can be used as the downstream boundary condition.

With the boundary information specified, the solution for v and y at a sufficient number of intermediate points and at the downstream boundary is obtained at the intersection of forward and backward characteristics and at the intersection of forward characteristics and the downstream boundary, respectively. The number of solution points must be sufficient to adequately describe the movement of a flood wave downstream and is determined by the number of characteristics inscribed on the (x,t) plane. Usually, the availability of data limits the number of characteristics; however, it should be noted that a better solution generally is obtained when more characteristic curves are involved. Mathematically, a complete solution is obtained if all the boundary information is utilized.

The usual procedure for solving Eqs. 5.5 and 5.6 simultaneously is shown in Figure 5.2. Consider the points numbered 2, 5, 6, 10, 11 and 12. There is both a forward and a backward characteristic emanating from points 2, 5 and 10. The intersection of the forward characteristic out of point 10 with the backward characteristic out of point 5 specifies the conditions (values of v and y) at point 11. Similarly, point 6 conditions are determined at the intersection of the respective character-istics out of points 5 and 2. The forward characteristic out of point 10 is continued downstream in time and space until it intersects with the backward characteristic out of point 2, thereby determining v and y at point 12. The procedure continues for all forward and backward characteristics until they intersect either the downstream or upstream boundary.

At each intersection there are four unknowns x, t, v, and y. These are uniquely determined by the simultaneous solution of the four equations given by Eqs. 5.5 and 5.6. At the downstream boundary x is given and is no longer an unknown. The other three unknowns are determined by the simultaneous solution of Eqs. 5.5a and 5.6a and rating

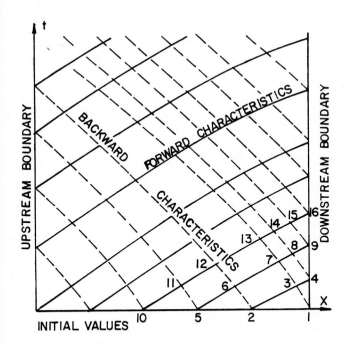

Fig. 5.2   Characteristics in (x,t) Plane

curves that relate v and y.

A method for solving the characteristic equations according to this procedure is outlined as follows. With reference to Figure 5.3, it is assumed the values for v and y are known at L and R and are desired at M. Eqs. 5.5 are approximated as

$$X_M = X_L + (t_M - t_L) (v + c)_L \tag{5.7}$$

and
$$X_M = X_R + (t_M - t_R) (v - c)_R \tag{5.8}$$

These two equations can be easily solved for the two unknowns X_M and t_M. Once these are known Eqs. 5.6 are solved by the same approach.

$$(v + 2c)_M = (v+2c)_L + (t_M - t_L)\lambda_L \tag{5.9}$$

and
$$(v-2c)_M = (v-2c)_R + (t_M - t_R)\lambda_R \tag{5.10}$$

where
$$\lambda_L = g(S_o - S_f)_L - (v-c)_L \frac{q_i}{A_L} \tag{5.11}$$

$$\lambda_R = g(S_o - S_f)_R - (v-c)_R \frac{q_i}{A_R} \tag{5.12}$$

and
$$c = \sqrt{gy} \tag{5.13}$$

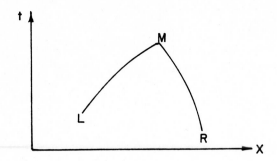

Fig. 5.3   Characteristic Solution for Point M

The method outlined by Eqs. 5.7 through 5.13 is linear and can be solved for v_M, c_M hence y_M. The boundary condition, inflow, initial values and downstream rating curve must be known. A more stable and accurate solution can be obtained with a nonlinear formulation (Overton and Meadows, 1976; and Mahmood and Yevjevich, 1975).

When solving problems using the method of characteristics, check whether the flow is subcritical or supercritical. When the flow is subcritical, $v < c$ and the forward characteristic has a positive slope dx/dt in the (x, t) plane while the backward characteristic has a negative slope, as shown in Figure 5.4a. When, however, the flow is supercritical, $v > c$ and both characteristics have a positive slope in the (x, t) plane, Figure 5.4b.

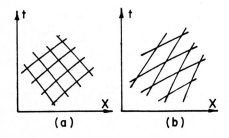

Fig. 5.4   Characteristic Lines for Subcritical and Supercritical Flows

FINITE DIFFERENCE METHODS

Finite differencing involves replacing the continuous derivative terms with approximate finite difference quotients, thereby transforming the set of differential equations into a set of either linear or non-linear algebraic equations which can be solved more readily. These algebraic equations relate unknown dependent variable values at nodal points on a finite grid overlaying the continuous solution domain to known initial values and boundary conditions. Solution of these equations is either direct or through a root determining scheme such as the Newton-Raphson Method. In either case, depending on the manner in which the replacements are made, matrix techniques may also be required.

Difference Quotients

Finite difference quotients are obtained by dividing the difference between two values of a function by the corresponding two values of the independent variable. For the case of a function of a single variable, e.g. $f(x)$, the difference quotient is given by

$$\frac{f(x+\Delta x) - f(x)}{\Delta x}$$

The limiting value, as $\Delta x \to 0$, is the definition of the derivative

$$\frac{df(x)}{dx} = \lim_{\Delta x \to 0} \frac{f(x+\Delta x) - f(x)}{\Delta x} \qquad (5.14)$$

Thus the finite difference quotient is an approximation to the continuous derivative as long as Δx is kept small.

Several difference quotients can be defined to approximate partial derivatives. To illustrate some of them, consider a function of two independent variables, say $U(x,t)$. With reference to the finite difference grid in Figure 5.5, the most commonly used difference quotients are defined as follows. The forward difference approximation to the first partial derivative for U with respect to x is

$$\frac{\partial U}{\partial x} = \frac{U(x+\Delta x, t) - U(x,t)}{\Delta x} \qquad (5.15)$$

Physically, one can think of an observer standing at the point (x,t), looking ahead (forward) to the point $(x+\Delta x, t)$, and using the elevation (function value) difference between the two points divided by the distance to evaluate the slope (value of the derivative). The backward difference approximation is

$$\frac{\partial U}{\partial x} = \frac{U(x,t) - U(x-\Delta x, t)}{\Delta x} \qquad (5.16)$$

The centered (or central) difference approximation is

$$\frac{\partial U}{\partial x} = \frac{U(x+\Delta x, t) - U(x-\Delta x, t)}{2\Delta x} \qquad (5.17)$$

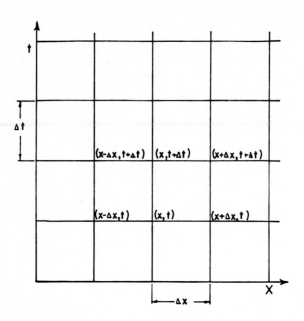

Fig. 5.5   Finite Difference Grid for x,t Solution Domain

NUMERICAL SOLUTION

There are two basic finite difference schemes used in solving the
streamflow routing equations. They are the explicit and implicit schemes.
Explicit schemes utilize initial value and left hand side (upstream)
boundary information and solve for the remaining grid points one at a
time. They are subject to stability limitations on the allowable grid
interval size which means explicit schemes typically have large data
requirements. However, explicit methods often result in linear algebraic
equations from which the unknowns can be evaluated directly without
iterative computations. Implicit schemes utilize initial value and both
left and right hand side boundary information, and solve for the
unknown grid points at the next time level simultaneously. Therefore,
implicit schemes often require matrix techniques. Implicit methods
typically involve nonlinear algebraic finite difference equations whereby
the solution is attained by iteration. Both schemes can be and have been
used in solving the governing equations for overland and open channel
flow.

Most existing methods for numerical solution of equations can be classified into the following groups:

(a) Explicit finite difference methods

(b) Implicit finite difference methods

(c) Finite element methods

The use of the first two methods was summarised by Liggett and Woolhiser (1967). They reviewed different explicit finite difference schemes. The schemes were:

a) method of characteristics

b) unstable method

c) diffusion method

d) Lax–Wendroff method

e) leap–frog method

The method of characteristics uses an irregular grid following the characteristic curves while the others use a rectangular grid for the solution of the equations.

The method of characteristics employs the fact that flow conforms to certain relationships along characteristic curves and therefore the solution is performed along the characteristic curves. The main advantages of the characteristic method is that it is accurate and fast. It is the most acurate method for the same initial point spacing of all methods. Its accuracy is a consequence of following the characteristic curves which describe the path of the disturbances in the flow. It also covers the x – t plane faster than any other method with the same initial point spacing. The main disadvantage of the method of characteristic is that data at intermediate points in the x – t plane is difficult to obtain in an acceptable form, requiring tedious interpolation techniques. If the method is applied to a two-dimensional problem the use of the characteristic method becomes even more difficult. More recently more elaborate methods of characteristics were developed. Abbott and Verwey (1970) used a four-point method of characteristics, i.e. utilising three different points in fixing the properties of a fourth point. This solution could only be used with the dynamic equations as the kinematic equations do not have negative characteristics required for this method.

The implicit method of solution involves simultaneous solution of all the flow properties by solving a matrix; its main advantage is that the ratio of space to time interval, $\Delta x / \Delta t$, is not governed by any stability criteria and the method is considered to be stable for any choice of Δx and Δt. Most previous investigators considered this to be an advantage. Liggett and Woolhiser (1967) report, however, that they were unable to make practical use of this 'advantage'. If they increased $\Delta x / \Delta t$ ratio more than would be allowed for in an explicit finite difference scheme, inaccuracy resulted and sometimes stability problems occurred. They suggest that the implicit methods seemed to be more advantageous when dealing with river problems but pointed out that attention should be paid to the accuracy of the results obtained.

Only a few investigators have used finite element methods in solving the St. Venant equations. The main reason for not being used is that finite element programs are expensive to run and accuracy and stability criteria can become tedious to apply.

Explicit finite difference schemes have been widely used in the past for the solution of the one-dimensional St. Venant equations. They differ from each other in the way they define their discharge and depth gradients, but they all express the flow properties at a certain time as a function of the flow properties at a previous time thus permitting an explicit solution. They are simple to use as they use a fixed regular grid and it is easier to follow the variation of the flow properties along the catchment as the solution is performed explicitly. They have been found to be accurate and economical when properly used. The main problems accompanying the choice and the use of an explicit finite difference scheme are, however, those of accuracy and stability. Choosing the most proper scheme and using it accordingly is, therefore, important in obtaining stable and accurate results.

The main explicit finite difference schemes which have been used previously are summarised in Figure 5.6 in terms of the points used at a time interval to propagate information at the next time interval.

The properties of the different schemes are summarized by Liggett and Woolhiser (1967). The unstable method was found to be unreliable while the rest showed signs of instability when used in certain cases. The Lax-Wendroff scheme tended to dampen out instabilities and produce better results.

Various other investigators were faced with similar problems when using such different schemes for the solution of the kinematic equations. Constantinides (1982), however, argued that as the nature of information propagation for the kinematic equations differs from that of the St. Venant equations alternative difference schemes had to be developed. Furthermore, he argued that the scheme to be used should propagate numerically, information in a similar manner as suggested by the kinematic characteristic equations. Using this he developed a scheme shown to be accurate, stable and fast   (Table 5.1, p. 103).

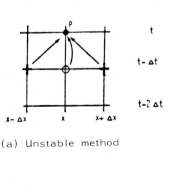

(a) Unstable method

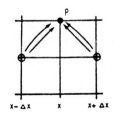

(b) Diffusing method

(d) Lax Wendroff method

Uses for diffusing scheme
for the first time interval
and the leap-frog scheme
for subsequent time intervals

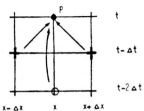

(c) Leap-frog method

p : Point where flow properties will be calculated (x,t)
+ : Points used for defining discharge gradients
o : Points used in the depth gradient definition
→ : Direction at which information is propagated
    for discharge
⤳ : Direction at which information is propagated
    for depth

Fig. 5.6   Explicit finite difference scheme used in the solution of the
one-dimensional St. Venant equations.

Explicit Scheme

The application of the explicit method to the unsteady flow equations
is primarily the outcome of pioneering work by J.J. Stoker; a complete
description is found in Isaacson, et al. (1956). The explicit scheme

shown here is from that report. A rectangular channel with no lateral inflow is assumed.

A network of node points is shown in Figure 5.7 for solving the governing equations using the explicit method. The variables are known at points L, M and R, and are to be determined for point P. Using a centered difference quotient to approximate the spatial derivatives and a forward difference quotient to approximate the temporal derivatives, the following approximations are made at point M:

$$v(M) = \frac{v(R) + v(L)}{2} \tag{5.18}$$

$$\frac{\partial A(M)}{\partial t} = \frac{A(P) - A(M)}{\Delta t} \tag{5.19}$$

$$\frac{\partial v(M)}{\partial x} = \frac{v(R) - v(L)}{2\Delta x} \tag{5.20}$$

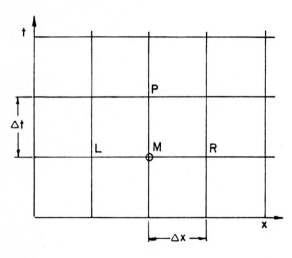

Fig. 5.7   Network of Points for Explicit Method

Similar approximations are made to the other derivative terms. When these approximations are inserted into Eqs. 5.1 and 5.2, $v(P)$ and $y(P)$ can be solved directly as

$$v(P) = \frac{v(R)+v(L)}{2} + \frac{\Delta t}{2\Delta x} \{\frac{v(L)^2 - v(R)^2}{2} + g[y(L)-y(R)]$$

$$+ g\Delta x [2S_o - S_f(R) - S_f(L)] \} \tag{5.21}$$

and

$$y(P) = \frac{1}{2}\{y(R)+y(L) + \frac{\Delta t}{\Delta x}[y(L)v(L)-y(R)v(R)]\} \tag{5.22}$$

The solution procedure is to use the information at time level t and solve for the unknowns at each of the grid points at time level t + Δt. Once this row of values has been determined, advance the computations to time level t + 2Δt. The values at time level t + Δt become the initial values for determining the unknowns at this advanced time level. The solution proceeds in this fashion until all the grid points in the solution domain have been determined.

To ensure stability, the grid sizes Δx and Δt are chosen to satisfy the constraint

$$\Delta t \leq \frac{\Delta x}{|v + c|} \qquad\qquad v + c \leq \frac{\Delta x}{\Delta t} \qquad\qquad (5.23)$$

This criterion for computational step sizes, known as the Courant condition, insures that the time increment is selected such that the node point P lies within the area bounded by the forward and backward characteristics generated from node points L and R. As discussed previously, this ensures that point P is within solution zone A and can be fully determined using only the initial value information contained along the line from L to R.

Implicit Scheme

A network of node points is shown in Figure 5.8 for solving the unsteady flow equations using an implicit method. the centered four point difference scheme is illustrated (Amein and Fang, 1969).

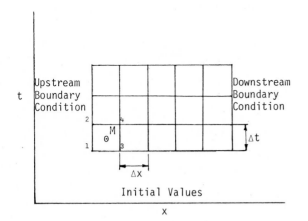

Fig. 5.8   Network of Points for Implicit Method

The following approximations to the derivative terms are made:

$$\frac{\partial Q}{\partial x} = \frac{Q(4) + Q(3) - Q(2) - (Q(1)}{2\Delta x} \qquad (5.24)$$

$$\frac{\partial A}{\partial t} = \frac{A(4) + A(2) - A(3) - A(1)}{2\Delta t} \qquad (5.25)$$

$$\frac{\partial y}{\partial x} = \frac{y(4) + y(3) - y(2) - y(1)}{2\Delta x} \qquad (5.26)$$

$$\frac{\partial v}{\partial x} = \frac{v(4) + v(3) - v(2) - v(1)}{2\Delta x} \qquad (5.27)$$

$$\frac{\partial v}{\partial t} = \frac{v(4) + v(2) - v(3) - v(1)}{2\Delta t} \qquad (5.28)$$

$$S_f = \frac{1}{4} \left[S_f(1) + S_f(2) + S_f(3) + S_f(4) \right] \qquad (5.29)$$

$$q = \frac{1}{4} \left[q(1) + q(2) + q(3) + q(4) \right] \qquad (5.30)$$

These approximations are used to replace the respective terms in Eqs. 5.1 and 5.2. Hydraulic variables at node points 1, 2 and 3 are known from boundary conditions and initial values, hence the unknowns are $Q(4)$, $v(4)$, $y(4)$, $A(4)$ and $S_f(4)$. Since $y(4)$ and $A(4)$ are related by the cross-sectional geometry and $Q(4)$, $v(4)$ and $A(4)$ are related by continuity, 'there are actually three unknowns and two equations. Since there is the need for another equation, the difference scheme is written for all of the distance steps at given time level until the downstream boundary is reached. In Figure 5.8 there are 12 grid boxes, meaning there will be 24 equations to be written but there will be 27 unknowns. The three additional equations are specified by the downstream boundary condition which most often is a rating curve between discharge and area (depth).

The resulting set of algebraic finite difference equations is non-linear and must be solved using an iterative root-finding scheme. Amein and Fang (1969) found that the Newton scheme could be used to linearize the equations which they then solved using matrix techniques.

The solution procedure is to solve for all the unknowns at one advanced time level before proceeding to the next. All values are determined simultaneously, and must satisfy all boundary conditions. Therefore, this method avoids the stability requirements of the explicit method meaning that larger x and t grid interval sizes can be used which requires less input data.

ACCURACY AND STABILITY OF NUMERICAL SCHEMES

There are two approximations in numerical modelling. One needs to ask the questions: "How well is the natural system modelled by the differential equations?", and, "How well is the solution to the differential equations represented by the computational algorithm?". In the analysis here more attention is paid to the second question. The first question can only be answered by studying the behaviour of the natural system and comparing it to the equations applied to it. Therefore it will be assumed here that the differential equations approximate the system well despite the fact that it has been noticed that this is not necessarily the case. Abbott (1974) noticed that a difference scheme considerably different from the differential equations used to describe a system, can yield more accurate results than a difference scheme similar to the differential equations when compared with experimental results.

There are three possible sources of error associated with finite difference solutions to partial differential equations. It is important that one understands these sources, their consequence if not controlled, and means for controlling them. These three sources of error are: truncation, discretization, and round-off. Truncation error occurs when a derivative is replaced with a finite difference quotient; discretization error is due to the replacement of a continuous model (function) with a discrete model; and round-off error is essentially machine error in that the algebraic finite difference equations are not always solved exactly.

For finite difference solutions to be accurate, they must be consistent and stable. Consistency simply means that the truncation errors tend to zero as Δx and $\Delta t \to 0$, i.e., as Δx and $\Delta t \to 0$ the finite difference equation becomes the original differential equation. This is examined in the following paragraphs. Stability implies the controlled growth of round-off error. Stability considerations apply principally to explicit schemes to be discussed later. Any numerical scheme that allows the growth of error, eventually "swamping" the true solution, is unstable.

Generally, to ensure stability requires that limits be placed on the allowable sizes for Δx and Δt. The criterion for establishing the allowable sizes is that they be chosen such that the forward and backward characteristics will not travel the distance Δx in the time interval Δt. This insures that the solution at the advanced point in time can be fully determined from available initial value information; i.e. the grid point being solved is in solution Zone A. Generally, if a numerical scheme is both consistent and stable, its solution will be convergent (accurate) with the solution of the partial differential equation.

The truncation error is examined with a Taylor's series expansion for $U(x,t)$ at the point $(x,0)$, i.e. time is held constant.

$$U(x+\Delta x,t) = U(x,t) + \Delta x \frac{\partial U}{\partial x} + \frac{\Delta x^2}{2!} \frac{\partial^2 U}{\partial x^2} + \cdots \qquad (5.31)$$

where the derivatives are evaluated at x,t. Dividing Eq. 5.31 by Δx, and rearranging, gives the series equivalent to the forward difference quotients, Eq. 5.8

$$\frac{U(x+\Delta x,t) - U(x,t)}{\Delta x} = \frac{\partial U}{\partial x} + \frac{\Delta x^2}{2!} \frac{\partial^2 U}{\partial x^2} + \cdots \qquad (5.32)$$

which shows that replacing $\partial U/\partial x$ with the forward difference quotient introduces an error of approximation equal to those terms on the right hand side of Eq. 5.32 after $\partial U/\partial x$. This error is proportional to the first power of Δx; we call this first order error (or approximation). Similarly, it can be shown that the backward difference quotient has first order error, and the centered difference has second order error.

Consider the following partial differential equation

$$\frac{\partial Q}{\partial t} + c \frac{\partial Q}{\partial x} = 0 \qquad (5.33)$$

One finite difference approximation to this equation is

$$\frac{Q(x+\Delta x,t+\Delta t) + Q(x,t+\Delta t) - Q(x+\Delta x,t) - Q(x,t)}{2\Delta t}$$

$$+ c \frac{Q(x+\Delta x,t+\Delta t) - Q(x,t+\Delta t)}{\Delta x} = 0 \qquad (5.34)$$

Examination of the Taylor's series residuals reveals the absolute value of the truncation error is

$$\text{Error} = \frac{v^2 \Delta t}{2} \cdot \frac{\partial^2 Q}{\partial x^2} + O(\Delta x^2, \Delta t^2) \qquad (5.35)$$

where the last term indicates a second order of approximation. On inspection it appears that Eq. 5.34 is consistent with Eq. 5.33 as $\Delta t \rightarrow 0$. However, for this particular solution, stability considerations require that

$$c \leq \frac{\Delta x}{\Delta t} \qquad (5.36)$$

Substituting this inequality into Eq. 5.34 transforms the error term into

$$\text{Error} \leq \frac{\Delta x^2}{2\Delta t} \frac{\partial^2 Q}{\partial x^2} + 0(x^2, t^2) \tag{5.37}$$

which indicates a small error term, but one that can become significant if $\Delta t \rightarrow 0$ faster than $\Delta x^2 \rightarrow 0$. Since Δx and Δt are finite and are not approximately zero, Eq. 5.34 approximates Eq. 5.33 with second order accuracy but with a term introducing artificial (numerical) dispersion.

This example was chosen because it illustrates how kinematic models can simulate a dispersing hydrograph. Eq. 5.33 is merely the kinematic wave equation for no lateral inflow which, theoretically, cannot predict hydrograph dispersion. Eq. 5.34 is one of the finite difference models used to solve the kinematic model. Because of the presence of the truncation error, it simulates a dispersing hydrograph, thereby demonstrating that a numerical kinematic model can simulate a dispersing hydrograph.

Numerical dispersion or diffusion is the process in which the Error is formed. It is the development of the truncation error to the error through the numerical technique used.

Lax's (1954) theory, proved by Richtmyer and Morton (1967) states that for linear equations with constant coefficients operating on uniformly continuous initial and boundary data the following theorem holds. Given a properly posed initial-value problem and finite difference approximation to it that satisfies the consistency conditions, stability is the necessary and sufficient condition for convergence. This is however proved only for linear equations and according to Abbott (1979) it breaks down when there are discontinuities in flow.

Since one is dealing with non-linear partial differential equations (p.d.e's) there is no rigorous proof specifying stability criteria. For linear p.d.e's, however, stability analyses exist. Von Neuman (1949) was first to devise a powerful technique for determining stability criteria for linear p.d.e's. He made use of the fact that just about any function can be represented by a Fourier series. The linear stability analysis method essentially determines how the Fourier coefficients behave (grow, decay, or stay constant) with time for any term in the Fourier series. For stability to occur the ratio of a Fourier coefficient of any term at any time over the Fourier coefficient of the same term at a previous time must be less than one.

The effect of Δx and Δt on stability and accuracy are summarized in Figure 5.9. From Figure 5.9 one can deduce that the main criteria in the selection of Δx and Δt values for an explicit finite difference scheme are :

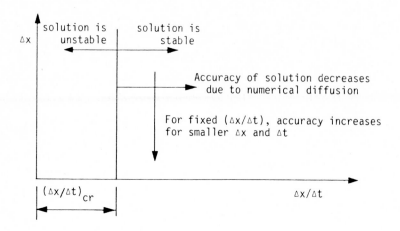

Fig. 5.9   Effect of value of Δx and Δt on stability and accuracy for and explicit finite difference scheme.

a)   that the scheme shall proceed under stable conditions

i.e. $\dfrac{\Delta x}{\Delta t} \geq \left(\dfrac{\Delta x}{\Delta t}\right)_{cr}$ (5.38)

b) $\dfrac{\Delta x}{\Delta t}$ shall be close to $\left(\dfrac{\Delta x}{\Delta t}\right)_{cr}$ to minimise diffusion errors and obtain optimal   accuracy.

c)   the difference scheme shall be convergent. This could be ascertained by running the scheme with different Δx's and Δt's and comparing with analytical results in a simple case.

$\left(\Delta x/\Delta t\right)_{cr}$ has been shown to be the speed of wave disturbance or information as it is propagated. This can be demonstrated by considering the manner in which information is propagated along the characteristic curves. For example, consider a central difference scheme, similar to the diffusion method, for solving the St. Venant equations. Let i represent a space interval, and k represent a time interval as shown in Figure 5.10. The point in question, i.e. where the flow properties are to be calculated, has the co-ordinates (i,k). Information about the flow properties is sought from the previous time interval. In Figure 5.10 (a) the true propagation speed is smaller than the numerical propagation speed while in Figure 5.10 (b) the converse is true. Numerical propagation lines are lines that have a slope $\Delta x/\Delta t$ in the x – t plane while true propagation lines have a slope dx/dt in the x – t plane. In Figure

5.10 (a) information is obtained within the i - l, i + l range by the true propagation lines. In Figure 5.10 information is sought by the true propagation lines outside the i - l, i + l range.

Since information outside this range is not propagated by the numerical scheme, it cannot be found and thus instability will result. A more detailed explanation is given by Stoker (1957).

For stability of an explicit finite difference scheme the following must therefore hold:

$$\frac{\Delta x}{\Delta t} \geq \frac{dx}{dt} \tag{5.39}$$

This is referred to as the "CFL condition" after Courant, Friedrichs and Lewy (1928), or simply the Courant criterion for stability.

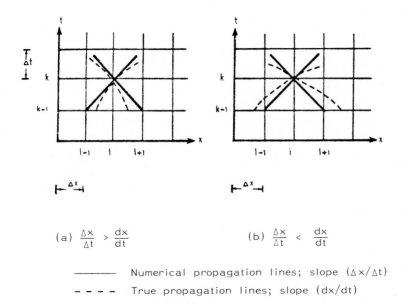

(a) $\frac{\Delta x}{\Delta t} > \frac{dx}{dt}$ (b) $\frac{\Delta x}{\Delta t} < \frac{dx}{dt}$

——————— Numerical propagation lines; slope $(\Delta x/\Delta t)$

- - - - True propagation lines; slope (dx/dt)

Fig. 5.10   Comparison of numerical and theoretical propagation of information in a central difference scheme

It has been noticed, however, that even if one satisfies the CFL conditions it is not necessarily true that the solution of the difference scheme is inherently stable (e.g. by LAX, 1954; Richtmyer and Morton, 1967; Abbott, 1974). There are two possibilities which could give rise to instability. There could be a physical discontinuity in the flow, e.g.

a bore or a hydraulic jump or parasitic waves could be generated within the difference scheme.

In terms of characteristics a physical discontinuity implies the intersection of two or more characteristics. Theoretically this results in different values of flow properties for a fixed place and time. In a difference scheme with a fixed grid this theoretical multivaluedness cannot be accounted for and in the solution is present in the form of oscillations. If the difference scheme tends to amplify these oscillations instability will occur. If however, these oscillations get damped stability will result and our scheme is referred to as a dissipative difference scheme.

The difference scheme being used can also cause oscillations called parasitic waves. It has been noticed (e.g. by Abbott, 1974) that the parasitic waves do not only occur when a physical discontinuity occurs but can arise out of the numerical procedure used. Therefore certain difference schemes have been found to produce parasitic waves while others do not when considering the same physical problem.

There are two ways these problems can be overcome. If a physical discontinuity exists, it can be located, the laws governing the discontinuity can be applied, and the laws governing continuous flow can be applied to each side.

It is also possible to adjust any difference scheme to dampen instead of amplify   parasitic waves. The solutions obtained from these "dissipative difference schemes", are called "weak solutions", as in this way stability is obtained at the loss of accuracy (see Lax, 1954). Abbott (1974) describes the dissipative schemes and the amount of accuracy lost extensively.

If one considers the method of setting up a dissipative scheme, one will also illustrate the principle of the weighted averages which is based on averaging flow properties at a certain time interval by linear interpolation according to where the characteristic curves intersect at t = constant line. Consider for example a backward difference scheme as shown in Fig. 5.11 and the way information about depth (y) is propagated. Depth at time t = k - 1 is taken to be as $\frac{1}{2}$ (1 - r)y_i^{k-1} + ry_{i-1}^{k-1} (see Figure 5.11).

Suppose now one wants to propagate the depth at point Q. Then interpolating linearly between points A and B one must use depth at Q at time t = k-1 as (1-r)y_i^{k-1} + ry_{i-1}^{k-1} , where r is the ratio of distance QB over distance AB in Figure 5.11.

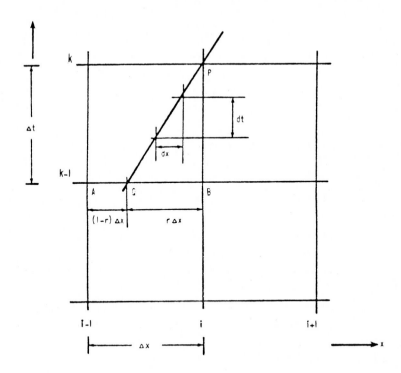

Fig. 5.11   The principle of weighted averages for information propagation in a backward difference scheme.

If one uses the fact that information is truly propagated as a speed
of $\frac{dx}{dt}$ then the slope of line QP, shown in Figure 5.11 should be the
value of $\frac{dx}{dt}$ at point Q (representing a point in space at a particular
time) denoted as $(dx/dt)_Q$

Strictly speaking the value of r should therefore be

$$r = (\frac{dx}{dt})_Q \,/\, \frac{\Delta x}{\Delta t} \tag{5.40}$$

A dissipative difference scheme is one as described above but with
r chosen in such a way as to dampen oscillations. The discrepancy
between r chosen and r in equation (5.40) will result in loss of accuracy
in the solution of the difference scheme.

EFFECT OF FRICTION

Because the friction term in the flow equation is non linear it makes solution of implicit type equations more difficult than without the friction term. A number of methods of accounting for the friction term was described by Cunge et al. (1980): The friction gradient is assumed to be of the form

$$S_f = Q|Q|/K^2 \tag{5.41}$$

where $K = AR^{2/3}/n$ (Manning, S.I. units) $\tag{5.42}$

$$R = A/P \tag{5.43}$$

If an explicit scheme is not acceptable, for instance if S_f is large compared with $\partial y/\partial x$, then some form of averaging of S_f in time is required. Cunge et al. suggest taking the average Q over the distance interval and squaring that, rather than the average of the squares of the Q's over the interval, i.e.

$$Q^2/K^2 = \frac{\theta}{4}\left[\left(\frac{Q}{K}\right)_j^{n+1} + \left(\frac{Q}{K}\right)_{j+1}^{n+1}\right]^2 + \frac{1-\theta}{4}\left(\frac{Q}{K}\right)_j^n + \left(\frac{Q}{K}\right)_{j+1}^n\right]^2 \tag{5.44}$$

An alternative which produces a linear equation and also yields the correct sign of Q was suggested by Stephenson (1984) for closed conduits:

$$Q|Q|/K^2 = \frac{\theta}{4}\left[Q_j^{n+1}|Q_j^n|/(K_j^n)^2 + Q_{j+1}^{n+1}|Q_{j+1}^n|/(K_{j+1}^n)^2\right]$$

$$+ \frac{1-\theta}{4}\left[(Q_j^n/K_j^n)^2 + (Q_{j+1}^n/K_{j+1}^n)^2\right] \tag{5.45}$$

Strelkoff (1970) indicates that the direct explicit scheme is inherently unstable. He indicates the Lax type scheme should satisfy the Courant criterion. For implicit schemes he suggests that to ensure stability in friction

$$\Delta t < \frac{K_o}{Ag/S_o} \tag{5.46}$$

where $K_o = A_o C \sqrt{R_o} = Q_o/\sqrt{S_o}$ $\tag{5.47}$

$$\therefore \quad \Delta t < \frac{V}{gS_f} \tag{5.48}$$

Wylie (1970) suggested that for a simple linear explicit system for open channels that for stability

$$\Delta t \leq (\Delta x/c)(1 - gS_f\Delta t/2V)^{1/2} \tag{5.49}$$

Even this does not guarantee stability according to Wylie.

CHOOSING AN EXPLICIT FINITE DIFFERENCE SCHEME FOR THE SOLUTION OF THE ONE-DIMENSIONAL KINEMATIC EQUATIONS

Constantinides (1982) used various schemes for solving the one-dimensional kinematic equations in an attempt to choose the most suitable scheme. The difference schemes mentioned earlier as well as new proposal schemes were used. The equations were solved for different problems which can also be solved with analytical methods. The analytical solutions were then compared with results from the numerical solutions. The suitability of the various difference schemes was then evaluated on the basis of accuracy and stability. The choice of a difference scheme was done by the process of elimination as more complicated problems were considered. A new proposal scheme, shown in Table 5.1 was found to yield extremely accurate results, to be stable as long as the Courant criterion is satisfied and to be fast and economic to run. The scheme is summarised in Table 5.1 by defining the discharge rate and depth at a time interval.

TABLE 5.1   Backward-central explicit difference schemes

Difference Scheme	Discharge Rate $\frac{\partial q}{\partial x}$ at $t = k - 1$	Depth y at $t = k - 1$
	$\dfrac{(q_i^{k-1} - q_{i-1}^{k-1})}{\Delta x}$	Y_i^{k-1}

INDEX

×   points where flow properties are to be calculated

+   points used for calculating discharge at time $t = k - 1$

0   points used for calculating depth at time $t = k - 1$

The explicit finite difference scheme shown in Table 5.1 although chosen by trial and error as being the most efficient scheme, becomes apparent when one considers the method of characteristics described earlier. The schemes propagate information downstream only as is suggested by the characteristic equation.

REFERENCES

Abbott, M.B., 1974. Continuous flows, discontinuous flows and numerical analysis. J. Hyd. Res., 12, No. 4.

Abbott, M.B., 1979. Computational hydraulics. Pitman Publ. Ltd. London

Abbott, M.B. and Verwey, A., 1970. Four-point method of characteristics. J. Hyd. Div., ASCE, HY12, Dec. 1970.

Amein, M. and Fang, C.S. (1969), Streamflow routing-with applications to North Carolina Rivers. Report No. 17, Water Resources Research Institute, University of North Carolina, Chapel Hill, North Carolina.

Constantinides, C.A., 1982. Two-dimensional kinematic modelling of the rainfall-runoff process. Water Systems Research Programme, Report 1/1982. Univ. of the Witwatersrand.

Courant, R., Friedrichs, K.O. and Lewy, H., 1928. Über die partiellen Differentialgleichungen der Mathematischen Physik, Math. Ann, 100.

Cunge, J.A., Holly, F.M. and Verwey, A., 1980. Practical Aspects of Computational River Hydraulics. Pitmans, Boston, 420 pp.

Isaacson, E., Stocker, J.J., and Troesch, B.A., 1956. Numerical solution of flood prediction and river regulation problems. Inst. Math. Sci. Report No. IMM-235, New York University, New York.

Lax, P.D., 1954. Weak solutions for non-linear hyperbolic equations and their numerical applications. Comm. Pure Appl. Math. 7.

Ligget, J.A. and Woolhiser, D.A., 1967. Difference solutions of the shallow water equation. J. Eng. Mech. Div. ASCE, April.

Lighthill, F.R.S. and Whitham, C.B., May 1955. On kinematic waves 1. Flood movement in long rivers. Proc. Roy. Soc. London, A, 229.

Mahmood, K. and Yevjevich, Eds., 1975 , Unsteady flow in open channels, Vols. I and II, Water Resources Publications, Fort Collins, Colorado.

Overton, D.E. and Meadows, M.E., 1976. Stormwater Modelling. Academic Press, New York.

Richtmyer, R.D. and Morton, K.W.,1967. Difference methods of initial value problems. 2nd Ed. Interscience, New York.

Stephenson, D. 1984. Pipeflow Analysis. Elsevier, Amsterdam, 274 p.

Stoker, J.J. 1957. Water Waves. Interscience Press, New York.

Strelkoff, T., 1970. Numerical solution of Saint-Venant equations. Proc. ASCE. J. Hydr. Div. 96(HY1), 223-252.

Von Neuman, J., 1963. Recent theories of turbulence. Collected Works (1949/1963) edited by A.H. Taub, 6, Pergamon, Oxford.

Wylie, E.B., Nov. 1970. Unsteady free-surface flow computations. Proc. ASCE, J. Hydr. Div., 96(HY11), 2241-2251.

CHAPTER 6

DIMENSIONLESS HYDROGRAPHS

UNIT HYDROGRAPHS

In the same way that the peak flow graphs in Chapter 3 can replace the Rational equation, so kinematic theory can be used to generate unit hydrographs for larger catchments. The simplifying assumptions in the Rational method and the peak flow charts are often inaccurate when it comes to larger catchments. An extension of the Rational method became necessary for large catchments and unit hydrograph theory was developed. The hydrograph shape was needed for routing too. An analogous procedure is developed below for selecting hydrographs for various catchment configurations. An advantage over the unit hydrograph methods is that the hydrographs here are dimensionless and allow for various simplified catchment configurations. This is offset by a slightly more complicated set of calculations. As with unit hydrograph procedures however, the catchment storm duration is selected by trial.

The dimensionless hydrographs presented below are synthesized for selected uniform storm durations. The catchments selected have varying shape and topography representing the majority of small catchments. The hydrographs, being dimensionless, are presented as functions of rainfall intensity and should therefore find international applicability. The user must select rainfall rates corresponding to desired return periods as well as initial abstraction and infiltration rates applicable to the catchment in question.

The hydrographs are intended for use by design engineers where not only the hydrograph peak flow rate but the shape of the hydrograph is important. The application to different catchments of varying shape and topography in developing the hydrographs makes their use more advantageous over other techniques, as explained below.

The lag effect due to overland flow length, surface roughness and slope is invariably included in the graphs presented. The result is a more realistic and effective hydrograph for the designer than is possible with previous methods. The effect of flow concentration in streams after flowing overland cannot be readily assessed using isochronal methods (or any other standard method). Neither can the effect of changing ground slope or converging flow which can all be accounted for with the kinematic models used here.

For peak discharge computation storms of duration smaller or equal to the time of equilibrium of the catchment are important, as a storm could produce maximum peak discharge off the catchment. Higher flood peaks may result from a shorter storm. The critical storm duration, i.e. the storm duration that will produce maximum peak discharge, will depend on two factors, these being the way the catchment responds to storms of duration less than the catchment's time of equilibrium, the rainfall characteristics and the retentive properties of the catchment's soils. Storms of durations longer than the catchment's time of equilibrium are also important, especially in cases where runoff volume is of importance.

Neither a single value of peak discharge rate nor total runoff volume are generally sufficient for all the purposes of the drainage engineer. The time the catchment takes to reach its peak discharge as well as the complete hydrograph shape are generally of prime importance. In cases where runoff hydrographs have to be combined from different catchments or are routed through hydraulic conduits, the complete runoff hydrograph shape is essential for accurate design.

The hydrograph shape is also important in designing hydraulic structures to cope with floods of higher return periods than those which they were designed to carry. The part of the hydrograph not carried by the hydraulic conduit structure, if known, can be diverted by suitable means, while, its backwater effects upstream and the force on the structure could also be evaluated.

The volume under the hydrograph is of particular importance when detention or retention storage are contemplated. The routing effect and peak flow attenuation are particularly sensitive to the hydrograph shape as opposed to the peak.

In general the dimensionless hydrographs should be of particular interest to the urban drainage engineer who will wish to study stormwater management and the effects of urbanisation – changing surface configuration, roughness and permeability on flow rates.

DEVELOPMENT AND USE OF GRAPHS

In developing runoff hydrographs for a catchment it is important to understand how the catchment will react to different storms. The volume of surface runoff is primarily a function of rainfall and infiltration characteristics, while the hydrograph shape is a function of catchment shape, roughness and topographical characteristics.

Computer models can account for any time and space variation of rainfall and catchment characteristics as described later. Their use entails substantial computer time and the model has to be used in conjunction with various storm inputs to ensure critical storm input. In this section, runoff hydrographs off catchments of fixed shapes and with spatially varied catchment characteristics are presented. The resulting hydrographs are dimensionless, i.e. in terms of catchment size and rainfall rate, allowing the use of different catchment dimensions and different roughness and catchment slope parameters. The design engineer can use these hydrographs for natural catchments which have similar shapes to the model catchments studied and where the roughness and slope characteristics are consistent. The design engineer still has to use his judgement in approximating catchment shapes and in averaging roughness and slope parameters.

The kinematic equations have been used to prepare the hydrographs presented by Constantinides and Stephenson (1982). Computer solution of the finite difference form of the equation of motion and the flow resistance equation was performed for numerous situations. With the use of dimensionless parameters the number of variables is reduced considerably and a few graphs present a range of hydrographs covering the range of parameters normally encountered.

Runoff hydrographs off three model catchments are presented, these being the following:

(a) A sloping plane catchment
(b) A converging surface catchment
(c) A V-shaped catchment with stream

Design hydrographs may be obtained by comparing dimensional runoff hydrographs for different storm durations, and selecting the one resulting in maximum flow rate (if the unattenuated peak is of concern) or greatest volume required to attenuate the flood if storage is to be designed, or any other relevant critical parameter.

List of Symbols

x space axis along overland plane (m or ft)

z space axis along channel (m or ft)

L_o length of overland plane (m or ft)

L_s length of channel or stream (m or ft)

S_o bed slope of overland plane

n_o roughness coefficient of overland planes

n_s roughness coefficient of channel or stream

θ angle describing converging surface catchment (radians)

r ratio describing converging surface catchment

w_o width of overland flow in converging surface catchment (m or ft)

H depth of channel (m or ft)

b width of channel (m or ft)

y_o depth of overland flow (m or ft)

q_o discharge per unit width of overland flow (m^2/s or ft^2/s)

y_s depth of channel flow (m or ft)

Q_s discharge of channel flow (m^3/s or ft^3/s)

Q_o discharge of converging surface (m^3/s or ft^3/s)

Kinematic equations

The one-dimensional kinematic equations for flow have already been presented and are merely stated here. They consist of the continuity equation and an equation relating hydraulic resistance to flow.

$$\frac{\partial Q}{\partial x} + \frac{\partial A}{\partial t} = q_L \qquad (6.1)$$

$$\text{and} \quad q = \alpha y^m \qquad (6.2)$$

where Q is the flow rate (m^3/s or ft^3/s), A is the cross sectional area (m^2 or ft^2), t is time (secs), x is the space axis (m or ft), q_L is lateral inflow per unit length along the x – axis (m^2/s or ft^2/s), q is the average discharge across a section per unit width (m^2/s or ft^2/s) and y is the depth of water (m or ft). α, m are coefficients dependent on surface roughness and bed slope.

EXCESS RAINFALL

In developing runoff hydrographs off the simple catchments already outlined, an excess rainfall distribution is required. In this case, excess rainfall intensity is assumed to be uniform in space, and constant during the storm and equal to a negative constant (being a constant infiltration rate) after the storm. Fig. 6.2 depicts the assumed excess rainfall input and   Fig. 6.1 shows the assumed rainfall input and loss distribution for obtaining the excess rainfall distribution shown in Fig. 6.2.

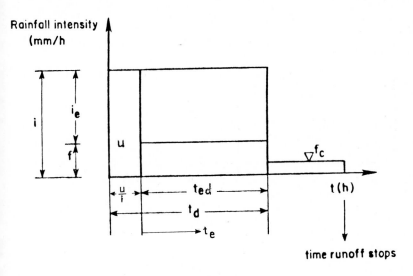

Fig. 6.1   Assumed rainfall input and distribution losses

In Figs. 6.1 and 6.2 i is rainfall intensity rate (mm/h or inches/h),i_e is excess rainfall intensity rate (mm/h or inches/h), t_d is storm duration (h), t_{ed} is excess rainfall duration (h), f_c is final infiltration rate (mm/h or inches/h),f is uniform infiltration rate (mm/h or inches/h) and u is initial abstraction (mm or inches).

The final infiltration rate, f_c is a function of soil type and vegetation cover or land use. The excess rainfall intensity, i_e, is a function of excess rainfall duration, t, which depends on local rainfall characteristics and on catchment soil and vegetation cover properties.

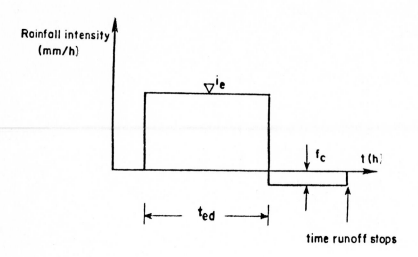

Fig. 6.2 Excess rainfall input

DIMENSIONLESS EQUATIONS

It is evident from kinematic theory that if any catchment is subjected to a constant excess rainfall intensity i_e for a period equal to or longer than its time of equilibrium it will produce a peak discharge equal to i_e multiplied by the area of the catchment. In deciding on dimensionless parameters to be used for developing runoff hydrographs it therefore seems logical to plot the ratio of discharge divided by excess rainfall intensity and area against a ratio of time divided by the time of equilibrium of a simple catchment, namely the sloping plane catchment.

Sloping Plane Catchment

For the sloping plane catchment depicted in Fig. 6.3 the continuity equation becomes:

$$\frac{\partial q_o}{\partial x} + \frac{\partial y_o}{\partial t} = i_e \text{ for } t \leq t_{ed} \tag{6.3a}$$

$$= -f_c \text{ for } t > t_{ed} \tag{6.3b}$$

The uniform flow equation can also be expressed as:

$$q_o = \alpha_o y_o^m \qquad (6.4)$$

where $\alpha_o = S_o^{\frac{1}{2}}/n_o$

and n_o, S_o are the Manning coefficients and bed slope respectively. Expressing y_o in terms of q_o from equation (6.4), differentiating with respect to t and substituting in equation (6.3) yields:

$$\frac{\partial q_o}{\partial x} + \frac{1}{\alpha_o^{1/m} \, m \, q_o^{1-1/m}} \frac{\partial q_o}{\partial t} = i_e \text{ for } t \leqslant t_{ed}$$

$$= -f_c \text{ for } t > t_{ed} \qquad (6.5)$$

The following dimensionless variables are then defined:

$$X = \frac{x}{L_o} \qquad (6.6)$$

$$P = \frac{q_o}{i_e L_o} \qquad (6.7)$$

$$T = \frac{mt_e}{t_{co}} \qquad (6.8)$$

$$T_D = \frac{mt_{ed}}{t_{co}} \qquad (6.9)$$

$$F = \frac{f_c}{i_e} \qquad (6.10)$$

where t_{co} is the time of concentration of a sloping plane in kinematic theory and is given by:

$$t_{co} = (\frac{L_o}{\alpha_o i_e^{m-1}})^{1/m} \qquad (6.11)$$

Substituting for q_o, x, t, t_{ed} and f_c in equation (6.5) and manipulating yields the following equation:

$$\frac{1}{P^{0.4}} \frac{\partial P}{\partial T} + \frac{\partial P}{\partial X} = 1 \text{ for } T \leqslant T_D$$

$$= -F \text{ for } T > T_D \qquad (6.12)$$

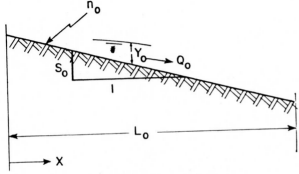

Fig. 6.3   Sloping plane catchment

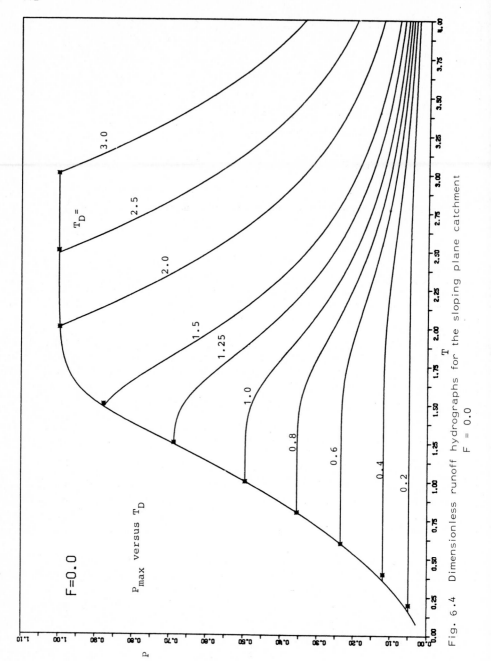

Fig. 6.4 Dimensionless runoff hydrographs for the sloping plane catchment
F = 0.0

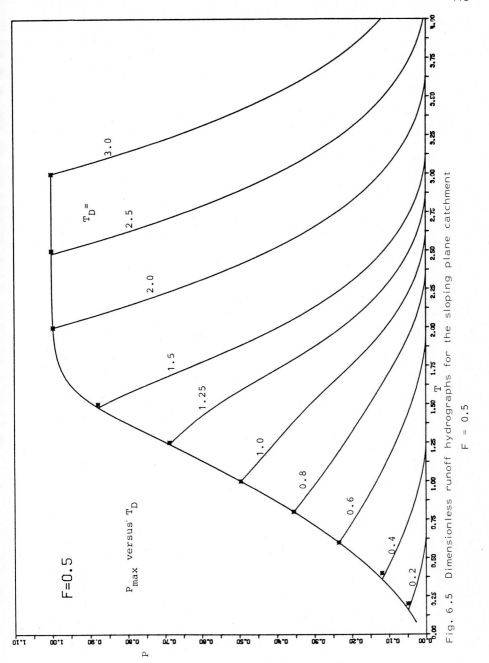

Fig. 6.5 Dimensionless runoff hydrographs for the sloping plane catchment

F = 0.5

where $(m-1)/m = 0.4$ for $m = 5/3$

Equation (6.12) is solved for flow P as a function of time ratio T at the outlet end of the catchment plane. This is repeated for different T_D values. Different plots are obtained for different F values in Figs. 6.4 and 6.5. The theory of Overton (1972) was also adapted to cascades of planes by Kibler and Woolhiser (1970).

Converging surface catchment

For the converging surface depicted in Fig. 6.6 the continuity eq. (6.1) becomes (Woolhiser, 1969):

$$\frac{\partial Q_o}{\partial x} + w_o \frac{\partial y_o}{\partial t} = w_o i_e \quad \text{for} \quad t \le t_{ed} \tag{6.13}$$

$$= -w_o f_c \quad \text{for} \quad t > t_{ed} \tag{6.14}$$

where $w_o = (L_o - x)\theta$

and $Q_o = w_o \alpha_o y_o^m$ \hfill (6.15)

Expressing y_o in terms of Q_o from equation (6.4), differentiating with respect to t and substituting in equation (6.13) yields:

$$\frac{\partial Q_o}{\partial x} + \frac{w_o^{1-1/m}}{m\alpha_o^{1/m}} \frac{1}{Q_o^{1-1/m}} \frac{\partial Q_o}{\partial t} = i_e w_o \quad \text{for} \quad t \le t_{ed}$$

$$= -w_o f_c \quad \text{for} \quad t > t_d \tag{6.16}$$

In addition to dimensionless variables defined in equations (6.8) to (6.10) the following dimensionless variables are defined (Singh, 1975):

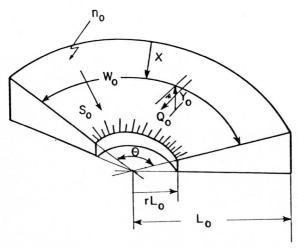

Fig. 6.6   Converging surface catchment

$$S = \frac{2Q_o}{i_e \theta L_o^2 (1-r^2)} \qquad (6.17)$$

$$X = \frac{x}{L_o(1-r)} \qquad (6.18)$$

where $(1-r^2)/2$ is the area of the catchment and r the ratio of bottom segment to the total catchment radius.

For the converging surface t_{co} is defined as the time of equilibrium for a sloping catchment of length $L_o(1-r)$, i.e.

$$t_{co} = \left[\frac{L_o(1-r)}{\alpha_o i_e^{m-1}}\right]^{1/m} \qquad (6.19)$$

Substituting for x, Q_o, w_o, t, t_{ed}, f_c and m in equation (6.16) and manipulating yields:

$$H \frac{\partial S}{\partial X} + \frac{H^{0.6}}{S^{0.4}} \frac{\partial S}{\partial T} \begin{array}{l} = 1 \text{ for } T \le T_D \\ = -F \text{ for } T > T_D \end{array} \qquad (6.20)$$

where $H = \dfrac{1 + r}{2\{1-X(1-r)\}}$ (6.21)

Equations (6.20) and (6.21) were solved numerically to give S as a function of T at the outlet for different T_D values. Plots are for various r and F values as presented in Figs. 6.7 and 6.8.

V-Shaped Catchment with Stream

In the V-shaped catchment (Fig. 6.9) the discharge from overland flow is used as input in the channel. Kinematic theory is used to route overland flow runoff through the channel. It is assumed that both overland flow planes are similar. From kinematic theory the continuity equation in the channel would be:

$$\frac{\partial Q_s}{\partial z} + b \frac{\partial y_s}{\partial t} = 2q_o L \qquad (6.22)$$

A basic assumption in equation (6.22) is that the natural depth of the channel is always greater than the water depth in the channel. Another assumption is that the channel area is small compared to the plane area. The uniform flow resistance equation for the channel may be written:

$$Q_s = b \alpha_s y_s^m \qquad (6.23)$$

Expressing y_s in terms of Q_s from (6.23), differentiating with respect to t and substituting into (6.22) yields:

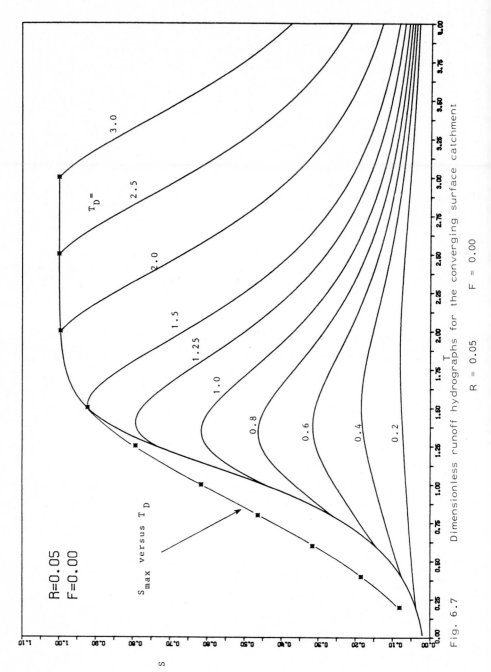

Fig. 6.7   Dimensionless runoff hydrographs for the converging surface catchment

R = 0.05   F = 0.00

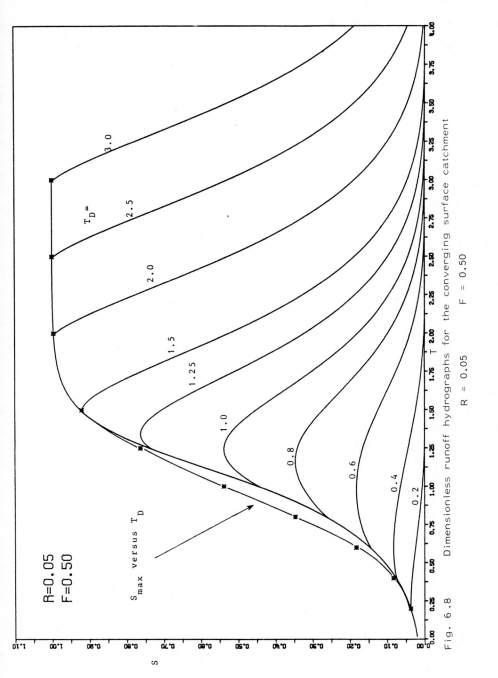

Fig. 6.8   Dimensionless runoff hydrographs for the converging surface catchment

R = 0.05   F = 0.50

$$\frac{\partial Q_s}{\partial z} + \frac{b^{1-1/m}}{m\alpha_s^{1/m}Q_s^{1-1/m}} \frac{\partial Q_s}{\partial t} = 2q_o \qquad (6.24)$$

In addition to the dimensionless variables defined in equations (5.6) and (6.10) the following dimensionless variables are defined:

$$Q = Q_s/2L_oL_si_e \qquad (6.25)$$

$$Z = z/L_s \qquad (6.26)$$

where t_{co} is the same as for the sloping plane, i.e. equation (6.11). Substituting for Q_s, z, t, q_o, and m in equation (6.24) and re-arranging yields:

$$\frac{G}{Q^{0.4}} \frac{\partial Q}{\partial T} + \frac{\partial Q}{\partial Z} = P \qquad (6.27)$$

where $G = (\frac{2L_s}{b\alpha_s})^{0.6} \frac{b\alpha_o^{0.6}}{2L_o} \qquad (6.28)$

Equation (6.12) is solved to yield P as a function of T at X = 1 for the planes. P is used as input in equation (6.27) to solve for Q as a function of T at the outlet for different values for F & G and the results appended at Figs. 6.10 and 6.13. The same problem was handled in a different' way by Wooding (1965).

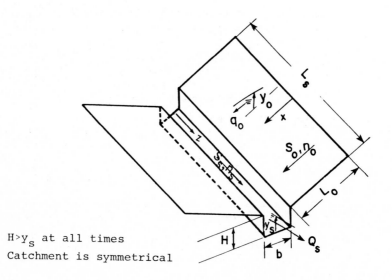

H>y_s at all times
Catchment is symmetrical

Fig. 6.9   V-shaped catchment with stream

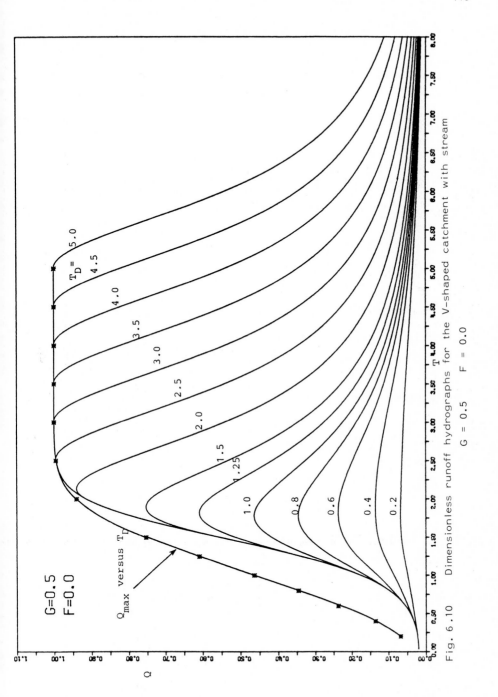

Fig. 6.10   Dimensionless runoff hydrographs for the V-shaped catchment with stream

G = 0.5   F = 0.0

120

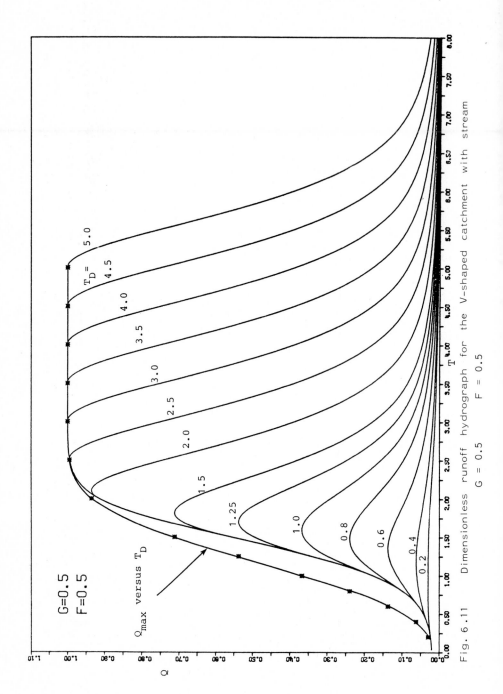

Fig. 6.11   Dimensionless runoff hydrograph for the V-shaped catchment with stream

G = 0.5   F = 0.5

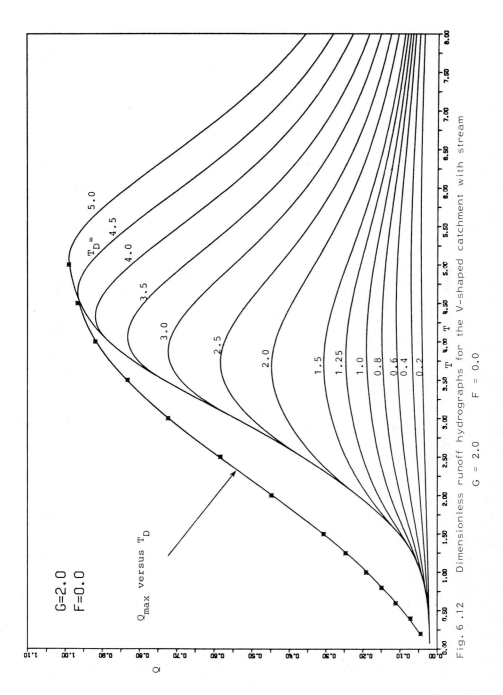

Fig. 6.12   Dimensionless runoff hydrographs for the V-shaped catchment with stream

G = 2.0   F = 0.0

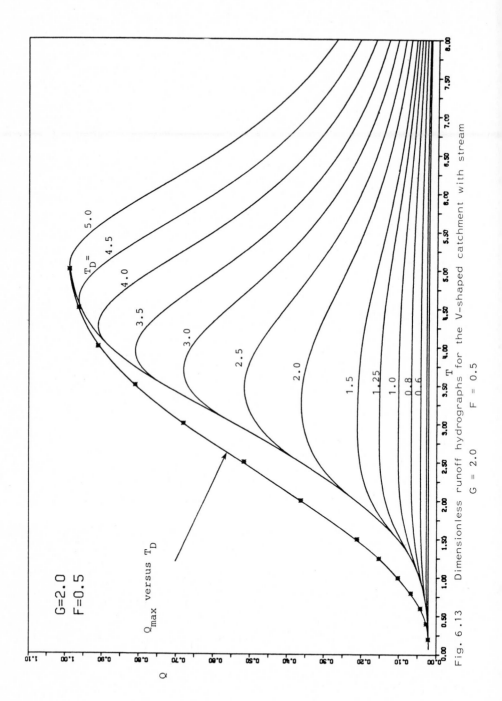

Fig. 6.13   Dimensionless runoff hydrographs for the V-shaped catchment with stream

G = 2.0   F = 0.5

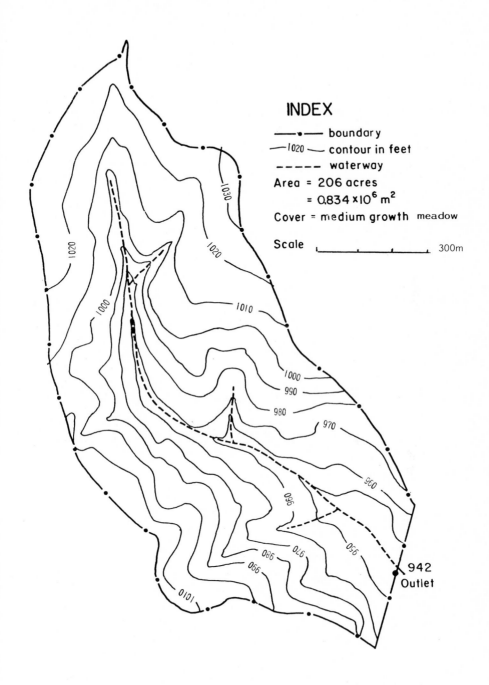

INDEX

—•— boundary
—¹⁰²⁰— contour in feet
----- waterway
Area = 206 acres
    = 0.834 × 10⁶ m²
Cover = medium growth meadow

Scale |————————| 300m

942
Outlet

Fig. 6.14   Example :   Catchment with stream

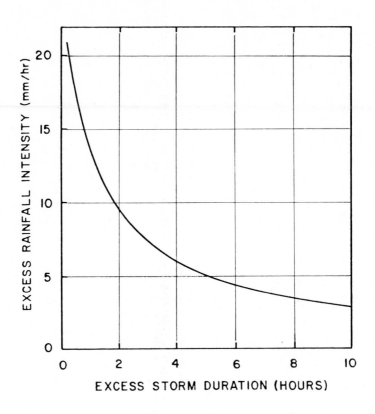

Fig. 6.15 Example : Excess intensity-duration relationship

TABLE 6.1 Example : Manning's roughness coefficients
and bed slopes

	Cover	Manning's n	Slope
Overland flow	Medium growth meadow	0.15	5%
Channel flow	Medium growth meadow	0.15	1.2%

USE OF DIMENSIONLESS HYDROGRAPHS

The procedure for using the dimensionless hydrographs is illustrated by means of an example.

Problem

Consider the natural catchment outlined in Fig. 6.14 and the 5 year recurrence interval excess IDF relationship shown in Fig. 6.15. Obtain the runoff hydrograph producing the worst peak discharge off the catchment. The excess IDF relationship given allows for the storm spatial distribution (which has been reduced from the point excess rainfall IDF relationship) and has been developed using local rainfall data and catchment characteristics. The average final infiltration rate of the soil (f_c) is 1.5 mm/h.

Solution

The natural catchment shown in Fig. 6.14 is approximated by a V-shaped catchment with stream. The main waterway in the catchment has a length of 1350 metres and subdivides the catchment approximately in the middle. The other waterways are minor and most of the catchment flow is in the form of overland flow flowing perpendicularly to the waterway. The waterway is assumed to be a rectangular channel 3m wide. The assumed V-shaped catchment with stream is illustrated in Fig. 6.16. Manning's roughness coefficients are shown in Table 6.1 while bed slopes are averaged using the contour lines from Fig. 6.14 and summarized in Table 6.1. Parameter G must be evaluated using (6.28):

$$G \quad = \quad \left[\frac{2(1350)}{\frac{3\sqrt{0.012}}{0.015}} \right]^{0.6} \quad \frac{3\left(\frac{\sqrt{0.05}}{0.15}\right)^{0.6}}{2(308.9)}$$

$$= \quad 0.44$$

Figs. 6.10 and 6.11 with G = 0.5 are used for choosing the critical runoff hydrographs. The infiltration parameter F is a function of the excess rainfall rate.

Table 6.2 shows the calculations in choosing a critical runoff hydrograph and dimensioning it. The table refers to Figure 6.10.

126

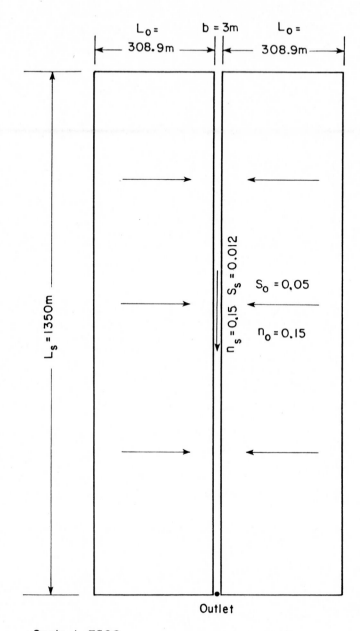

Scale 1 : 7500

Fig. 6.16   Example   : Assumed catchment

TABLE 6.2 Example : Choosing and dimensioning runoff hydrograph with maximum peak discharge

$$t_{co} = \left[\frac{L_o}{\alpha_o\, i_e^{m-1}}\right]^{1/m} \quad \text{(metric units) or } t_{co}\text{(hrs)} = \frac{L_o}{\alpha_o} = \frac{(3.6\times10^6)^{0.6}}{3600}\cdot\frac{1}{i_e^{0.4}} = \frac{2.858}{i_e^{0.4}} \quad (i_e \text{ in mm/hr})$$

$f_c = 1.5$ mm/hr $A = 0.834 \times 10^6$ m²

									Factors to dimension runoff hydrograph Multiply	
description / variable									horizontal axis	vertical axis
units	hours	mm/hr	-	hours	-	dimensionless hydrographs		mm/hr	m³/s	hours	m³/s
variable	t_{ed}	i_e	F	t_{co}	T_e	Q		$\dfrac{Q_s}{A}$	Q_s	$\dfrac{3}{5}t_{co}$	$\dfrac{i_e A}{3.6\times10^6}$
source	guess	excess IDF's	f_c/i_e	$2.858/i_e^{0.4}$	$\dfrac{5}{3}\dfrac{t_{ed}}{t_{co}}$	Fig	Value	$Q\cdot i_e$	$\dfrac{Q_s}{A}\times\dfrac{A}{3.6\times10^6}$		
	1.0	13.99	0.107	0.995	1.675	6.10	0.828	11.59			
	0.6	17.55	0.086	0.909	1.101	6.10	0.505	8.86			
	1.2	12.70	0.118	1.034	1.934	6.10	0.918	11.66	2.70	0.620	2.942
	1.4	11.63	0.129	1.071	2.178	6.10	0.971	11.29			

Critical storm has an excess duration of 1.2 hours producing a discharge peak of 2.70 cumecs.

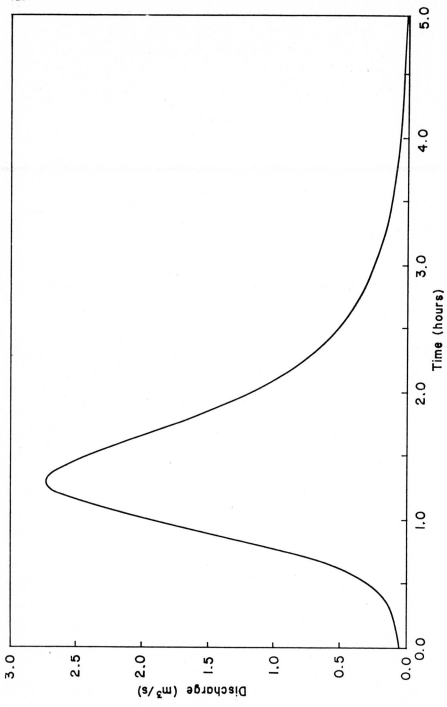

Fig. 6 .17   Example   :   Critical   runoff   hydrograph

As can be seen from Table 6.2 the storm producing the maximum peak discharge off the catchment has an excess storm duration of 1.2 hours and produces a peak discharge of 2.70 cumecs. The complete runoff hydrograph is obtained from Fig. 6.10 for a value of T_D = 1.93. The hydrograph is rendered dimensional by multiplying the two axes of Fig. 6.10 by the values given in Table 6.2 and is shown in Fig. 6.17.

REFERENCES

Constantinides, C.A. and Stephenson, D., 1982. Dimensionless hydrographs using kinematic theory, Water Systems Research Programme, Report 5/1982, University of the Witwatersrand.
Kibler, D.F. and Woolhiser, D.A., 1970. The kinematic cascade as a hydraulic model. Hydrol. paper 39, Colorado State University, Fort Collins.
Overton, D.E., 1972. Kinematic flow on long impermeable planes, Water Res. Bull. 8 (6).
Singh, V.P., 1975. Hydrid formulation of kinematic wave model of watershed runoff, J. Hydrol. 27.
Wooding, R.A., 1965. A hydraulic model for the catchment stream problem, II. Numerical Solutions. J. Hydrol. 3.
Woolhiser, D.A., 1969. Overland flow on a converging surface. Trans. Am. Soc. Agr. Engr. 12(4), 460-462.

CHAPTER 7

STORM DYNAMICS AND DISTRIBUTION

DESIGN PRACTICE

It is common practice to design stormwater systems for uniform intensity, uniformly distributed, stationary storms. Lack of data often makes any other basis for design difficult. There is little information available on instantaneous precipitation rates, storm cell size and cell movement. Time average precipitation rate or precipitation depth can be predicted from intensity-duration-frequency curves (e.g. Van Wyk and Midgley, 1966) or equations such as that of Bell (1969). The most common method of abstracting data from rainfall records is to select a duration and calculate the maximum storm precipitation in that period. The so-defined storm may include times of low rainfall intensity immediately preceding and succeeding a more intense precipitation rate.

Such simplifications in data render runoff calculation simplistic. Even when employing numerical models it is simplest to use a uniform intensity hyetograph for every point on the catchment. Although time varying storms are sometimes used, the precipitation pattern is seldom related to the maximum possible runoff rate.

Warnings have been made against simplification in rainfall patterns. For example, James and Scheckenberger (1983) indicated that storm movement can affect the runoff hydrograph significantly. Eagleson (1978) has expounded on the spatial variability of storms and Huff (1967) studied the time variability of storms.

Although much research has been done on storm variability, relatively little has been published on the resulting effects on runoff hydrographs (Stephenson, 1984). Research appears to have concentrated on models of particular (monitored) storms over particular catchments. The design engineer or hydrologist does not have sufficient guidance as to what storm pattern to design for. Presumably certain rainfall sequences, spatial variations and storm movement will result in a higher rate of runoff than other rainfall patterns for a particular catchment. Apart from an indication of what storm pattern produces the worst flood, one needs an indication of what storm pattern could be expected for the design catchment. Such data should be available on a frequency basis in order to estimate the likelihood of the worst hyetograph shape, spatial storm distribution and movements occurring. Although isolated catchments have been studied at many research centres considerably more information is required for the country as a whole. Analysis and use of such data in different combinations would require many trials before the worst storm patterns would emerge. An alternative approach is a deterministic one. Before calculating runoff, the analyst determines the following in order to select the correct design storm:

i)   The storm duration. For small catchments this is usually equated to the time of concentration of the catchment.
ii)   Variation in precipitation rate during the storm
iii)   Spatial distribution of the storm; and the
iv)   Direction and speed of movement of the storm.

The above information could be employed in numerical modelling of the design storm. Alternatively, for minor structures, simplistic methods such as the Rational method could be employed. Since data shortage often limits the accuracy of modelling, the latter, manual approach, is often sufficiently accurate. The guides presented below may assist both the modeller by providing information on which design storm would produce the highest runoff rate and the formula orientated solution by providing factors to account for storm variability.

STORM PATTERNS

Variation in rainfall intensity during a storm

In order to understand the reasons for and extent of variability (spatial and temporal) of rainfall, it is useful to describe the physical process of cloud formation and precipitation. Convective storm clouds originate from rising air masses. The size and shape of the rising air mass depends on the topography and the air masses will usually be of smaller scale than the air mass which has been brought by advection and which contains sufficient moisture for raindrops to precipitate. Mader (1979) concluded from radar observations of storms in South Africa that storm areas, duration and movement were related to mean 500 mb winds, thermal instability and wind shear.

Most recorded hyetographs indicate that rainfall intensity is highest somewhere in the middle of the storm duration. Huff (1967) presented extensive data on rainfall rates for storms of varying intensity indicating a time distribution somewhat between convex upward and triangular. In order to create a hyetograph which could be used for simple design of interconnecting stormwater conduits, Keifer and Chu (1957) proposed an exponential distribution termed the Chicago storm.

The position of the peak intensity could be varied and was observed to occur about 0.375 of the storm duration from the start.

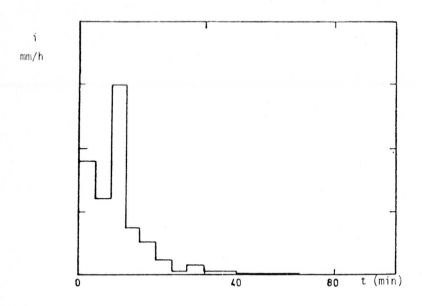

Fig. 7.1 Hyetograph with peak near beginning

Spatial distribution

The nature of storm cells within a potential rain area has been documented by many researchers e.g. Waymire and Gupta (1981). The persistence of storms observed in the northern hemisphere has not been found in countries south of the equator however (Carte, 1979). The larger air mass within which storm cells occur is referred to as the synoptic area (see Fig. 7.2). The synoptic area can last for 1 to 3 days and the size is generally greater than $10^4 km^2$. Within the synoptic area are large mesoscale areas (LMSA) of 10^3 to 10^4 km^2 which have a life of several hours. Sometimes small mesoscale areas (SMSA) of 10^2 to 10^3 km^2 can exist simultaneously. Within the mesoscale areas or sometimes on their own, convective cells, which are regions of cumulus convective precipitation, exist. These may have an area extent of 10 to 30 km^2 and have an average life of several minutes to half an hour. These cells are of concern to the hydrologist involved in stormwater design. By comparing the storm cell size with the catchment size he can decide whether the cell scale is significant in influencing spatial distribution over the catchment. There may be overlapping cells which could result in greater intensity of precipitation than for the single cells. Eagleson (1984) investigated the statistics of storm cell occurrences in a catchment and found the possibility of large storms can be computed assuming over-lapping small storms.

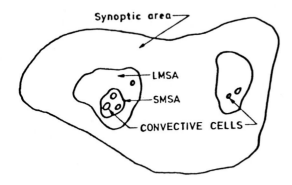

Fig. 7.2 Areal distribution of a convective storm

The shape of the storm cell has significance for catchments larger than the cell. Scheckenberger (1984) indicates that the cells are ellip-tical which may be related to storm movement. The rainfall intensity is highest at the centre and decreases outwards. The intensity has been shown to decrease exponentially, radially outwards from the focus, in various localities as in Fig. 7.3 (Wilson et al., 1979). Generally the variability in intensity does not necessarily cause higher runoff intensities but on small catchments near the centre of the cell the average precipi-tation can be higher than for a larger catchment, and as a rule, the rainfall depth increases the smaller the storm area.

Storm movement

Clouds generally travel with the wind at their elevation. As the rain falls it goes through lower speed wind movements so that the most significant speed is that of the clouds. The direction of lower winds can also differ from the general direction of movement of the upper strata. This may be the reason Changnon and Vogel (1981) observed slightly different directions for storm and cloud movements. Dixon (1977) analysed storm data and indicated storm cells have a circulation in addition to a general forward movement.

NUMERICAL MODELS

The effect of storm dynamics and distribution can be studied numerically and the results for simple plane catchments are presented below. The kinematic equations are employed in the numerical scheme. Although these solutions are no substitute for detailed catchment modelling when there are sufficient data, they do indicate which variables are likely to be the most important in storm dynamics. It must be pointed out that the following studies are simplified to the extent of assuming constant speed storms with unvarying spatial distribution. True storms are considerably more complex as explained in the above reference.

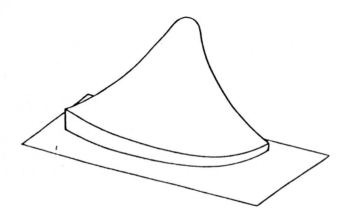

Fig. 7.3 Illustration of spatial distribution of precipitation intensity

Kinematic equations

The one-dimensional kinematic equations are for a simple plane catchment (Brakensiek, 1967):

The continuity equation; $\quad \dfrac{\partial y}{\partial t} + \dfrac{\partial q}{\partial x} = i_e \quad$ and

Flow resistance; $\quad q = \alpha y^m$

y is water depth on the plane, q is discharge rate per unit width of plane, i_e is excess rainfall rate, t is time, x is longitudinal distance down the plane, α is assumed a constant and m is a coefficient. Employing the Manning discharge equation in S.I. units $\alpha = \sqrt{(S_o)}/n$ where S_o is the slope of the plane, n is the Manning roughness coefficient, and m is 5/3.

The number of variables can be reduced to facilitate solution by re-writing the equations in terms of the following dimensionless variables:

$X = x/L$

$T = t/t_c$

$I = i_e/i_a$

$Q = q/i_a L$

where L is the length of overland flow, i_a is the time and space averaged excess rainfall rate and t_c is the time to equilibrium, or time of concentration, for an average excess rainfall i_a. Subscript c refers to time of concentration, d to storm duration, a to time and space average and p to peak. Then the following expression for t_c can be derived:

$$t_c = (L/\alpha i_e^{m-1})^{1/m}$$

In general the dimensionless variables are proportional to the dimensioned variables. Thus Q is the proportion of maximum flow at equilibrium. Substituting $y = (q/\alpha)^{1-m}$ from the resistance equation and for X, T, I and Q from the equations for the dimensionless terms, the following equation replaces the continuity equation.

$$\frac{\partial Q}{\partial T} = mQ^{1-1/m}(I - \frac{\partial Q}{\partial X})$$

This single equation can be solved for Q in steps of T and X for various distributions of I and m = 5/3.

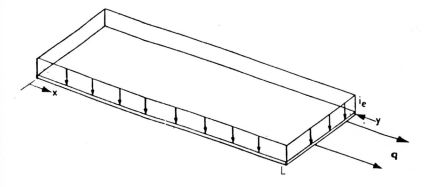

Fig. 7.4 Plane rectangular catchment studied with storm

Numerical Scheme

Although it appears a simple matter to replace differentials by finite difference, there can be problems of convergence and speed of solution unless the correct numerical scheme is employed. The simplest finite difference schemes are explicit, employing values of Q at a previous T to estimate new values at the next time T. This method is not recommended as it is often unstable when discontinuities in rainfall intensity occur. Upstream differences are usually taken in such schemes, as downstream effects cannot be propagated upstream according to Huggins and Burney (1982). It is also necessary to limit the value of $\Delta T/\Delta X$ to ensure stability.

Woolhiser (1977) documented various numerical schemes including very accurate methods such as Lax-Wendroff's. Brakensiek (1967) suggested 3 schemes: four point, implicit and explicit. His second scheme (implicit) is adopted here as it is accurate and rapid for the examples chosen.

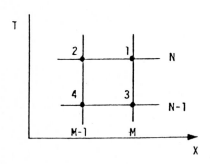

Fig. 7.5   X-T grid employed in numerical solution

Employing the notation in the grid in Fig. 7.5,

$$\frac{\partial Q}{\partial X} = \frac{Q_1 - Q_2}{\Delta X}$$

$$\frac{\partial Q}{\partial T} = \frac{Q_1 + Q_2 - Q_3 - Q_4}{2\Delta T}$$

Since $\partial Q/\partial T$ is not sensitive to $Q^{2/5}$, (the power is less than one), $Q^{2/5}$ is approximated by $((Q_3 + Q_4)/2)^{2/5}$, i.e. an explicit form is employed here or else the resulting equations would be difficult to solve. The finite difference approximation to the differential equation is thus:

$$\frac{Q_1 + Q_2 - Q_3 - Q_4}{2\Delta T} = \frac{5}{3}\left(\frac{Q_3 + Q_4}{2}\right)^{2/5}\frac{(Q_1 - Q_2)}{\Delta X}$$

solving for Q_1:

$$Q_1 = \frac{\frac{5}{3}\frac{(Q_3+Q_4)}{2}^{0.4}(1 + \frac{Q_2}{\Delta X}) + \frac{Q_3+Q_4-Q_2}{2\Delta T}}{\frac{1}{2\Delta T} - \frac{5/3}{\Delta X} \cdot (\frac{Q_3+Q_4}{2})^{0.4}}$$

Starting at the upstream end of the catchment where $Q_2 = 0$ and replacing Q_2 at the next point by Q_1 at the previous point, all the variables on the right hand side are known and one can solve for Q_1. The dimensionless time step used was 0.05. The difference for smaller time steps was found by trial to be unnoticeable.

SOLUTIONS FOR DYNAMIC STORMS

Time varying storms

One of the most frequently used simplifying assumptions, but a dangerous assumption, in many rainfall-runoff models is that of constant precipitation rate throughout the storm duration. The temporal variation of precipitation intensity for storms over Illinois was documented by Huff (1967) whose findings were often extrapolated to other regions. He suggested identifying the quartile of maximum precipitation and further employing probabilities of the rains occurring sooner or later than the median. Huff plotted his results as mass rainfall curves so it is not easy to discern the shape of the hyetographs unless his curves are differentiated with respect to time. In general they are found to be convex upwards. Apart from Keifer and Chu's (1957) synthetic hyetograph, evidence points to convex up hyetographs. The assumption of a triangular hyetograph is thus extreme as a real storm would tend to be less 'peaky' than a triangular one. The general triangular-shaped rainfall rate versus time relationship depicted in Fig. 7.6 is therefore studied. The time of the peak is varied between the start of the storm ($T_p = 0$) and the end ($T_p = 1$).

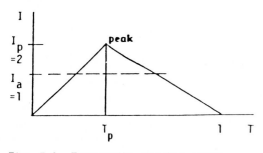

Fig. 7.6   Temporally varying storm

Simple models of hyetographs assume a single peak in rainfall intensity. Storms with multiple major peaks can be synthesized from overlapping compound storms. It is a single peak-storm which is considered here and the time of the peak intensity permitted to vary.

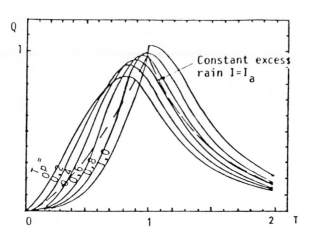

Fig. 7.7   Simulated dimensionless hydrographs caused by storms with time varying rainfall intensities (Fig. 7.6) but the same total precipitation

Design storms for flood estimation generally peak in intensity in the first half of the storm. This is an alleviating factor in peak runoff, as indicated in Fig. 7.7. That is a plot of hydrographs from the simple catchment depicted in Fig. 7.4 with various hyetographs imposed, i.e. a rectangular hyetograph and triangular hyetographs with various peak times were employed. The ordinate in Fig. 7.7 is the discharge rate expressed as a fraction of the mean excess precipitation rate, and the abscissa is time as a fraction of the time of concentration for a uniform storm with precipitation rate equal to the mean rate over the storm for each of the triangular hyetographs.

It will be observed from Fig. 7.7 that if the storm intensity peaks in the first part of its duration ($T_p \leqq 0.5$) the peak runoff is less than that for a uniform storm of the same average intensity. This holds for peaks up to 80% of the duration after commencement of rain. Only for the peak at the end of the storm (e.g. $T_p = 1.0$) does the peak runoff exceed that for a uniform intensity storm. Then the peak runoff is approximately 10% greater than for a uniform storm of the same duration.

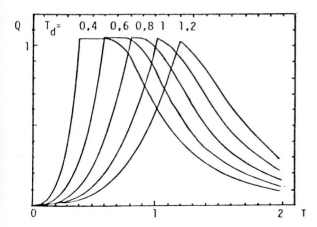

Fig. 7.8 Simulated dimensionless hydrographs caused by late peaking storms of constant volume and varying duration

If the storm duration is not equal to the time of concentration for a uniform storm however, the peak can be higher. Fig. 7.8 is for a storm of constant volume peaking at its termination (T_p = 1) and for durations represented by T_d = 0.4 to 1.2. These hydrographs are for storms of equal volume so that the shorter duration storms are of a higher intensity than longer duration storms. Depending on the IDF curve then a short duration storm may or may not result in a higher runoff rate than for one of duration equal to the concentration time of the catchment.

It should be recalled that all other hydrographs plotted are for a specified excess rate of precipitation. That is, if the hyetograph is uniform so are the abstractions. In practice, losses will be higher at the beginning of a storm, resulting in a late peak in excess rain even for a uniform precipitation rate. This has the same effect as a storm peaking in the latter part as it increases the peak runoff. The effect is compounded as a storm which peaks near the end will occur on a relatively saturated catchment so a greater proportion of the higher rate of rain will appear as runoff near the end. This tends to make the excess rain versus time graph concave upwards if the hyetograph was a straight-lined triangle. This effect is not modelled here but all the effects result in a higher peak than for a uniform input. Scheckenberger in fact indicates peaks up to 30% greater than for uniform storms due to the sum of these effects.

Spatial variations

It appears that areal distribution of the storm is less effective than temporal distribution in influencing peak runoff rate. Fig. 7.10 represents the simulated runoff from a 2-dimensional plane subjected to various distributions of a steady excess rain. The storm duration was made infinite in case the time to equilibrium exceeded the storm duration. The spatial (or longitudinal in this case) distribution was assumed triangular, the peak varying from the top to the bottom of the catchment as in Fig. 7.9.

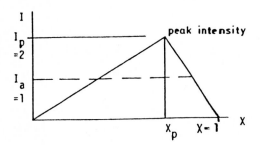

Fig. 7.9   Catchment with longitudinally varying storm

The same example would apply to a uniform intensity storm over a wedge-shaped catchment, the catchment width increasing linearly to X_p and then decreasing linearly towards the outlet where $X = 1$.

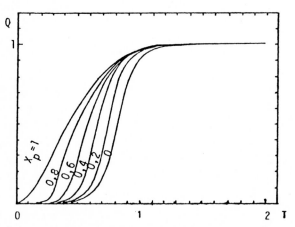

Fig. 7.10   Simulated dimensionless hydrographs caused by steady semi-infinite storms of varying distribution down catchment (Fig. 7.9).

Fig. 7.10 depicts the resulting simulated hydrographs which indicate that the runoff never exceeds that for a rectangular spatial distribution of rainfall. The resulting dimensionless time to equilibrium is
nearly unity for all cases, implying the same time of concentration holds
for uneven distribution as for uniform distribution of rain. There is
therefore not a chance of a shorter duration storm with a higher
intensity contributing to a greater peak than the uniform storm (unless
the intensity-duration curve is abnormally steep) since the time to equilibrium is not reduced relative to a uniform storm.

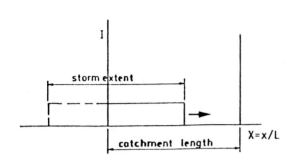

Fig. 7.11 Catchment with a storm moving down it

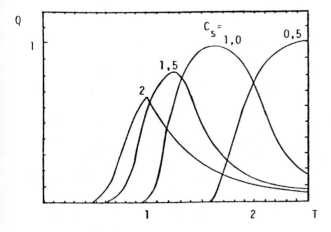

Fig. 7.12 Simulated dimensionless hydrographs caused by unit steady
uniform storms moving down catchment at different speeds
(see Fig. 7.11)

Moving storms

Fig. 7.12 represents simulated hydrographs from a storm with a constant precipitation rate and spatially uniform travelling down the catchment. The longitudinal extent of the storm cell is the same as the length of the catchment since in general smaller area storms are reputed to be more intense than larger cells. C is X/T_c or the speed divided by the rate of concentration. For slow storms $(C \leq 1)$ the dimensionless hydrograph peak is unity while for faster storms the peak is less. The faster storms do not fall on the catchment long enough to reach equilibrium.

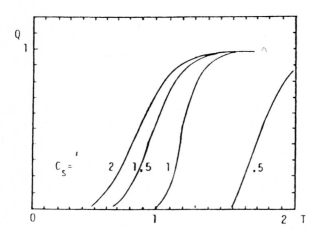

Fig. 7.13   Simulated dimensionless hydrographs caused by steady uniform semi-infinite storms moving down catchment at different speeds

Fig. 7.13 indicates there is also no increased peak for storms of semi-infinite longitudinal extent (never ending once they enter the catchment). All peaks converge on unity and there is no peak greater than unity. Thus movement does not appear to result in a hydrograph peak greater than for a stationary storm.

For storms of limited extent travelling up the catchment, the peak flow was observed to be less than for a stationary storm and the faster the speed of travel of the storm the smaller the peak runoff.

It has been demonstrated using numerical solutions to the kinematic equations for simple catchments that non-uniformity in rainfall intensity can affect peak runoff rates. Temporal variation in excess precipitation rate can increase runoff rate above that for a steady rate of rain. Since storms usually peak sometime after commencing and time diminishing abstractions tend to cause a later peak in excess rainfall rate, the assumption of steady rainfall can be dangerous as peak runoff is underestimated.

Uneven spatial distribution of a storm does not directly contribute to a higher peak runoff unless it results in a shorter duration storm being the design storm. Storm movement reduces the peak flow unless the movement is down-catchment, when this model shows no change in peak runoff rate. A smaller, more intense storm than the one to equilibrium for the catchment may however result in a higher peak runoff rate.

REFERENCES

Brakensiek, D.L., 1967. Kinematic flood routing. Trans Amer. Soc. Agric. Engs. 10(3) p 340-343.

Bell, F.C., 1969. Generalized rainfall-duration-frequency relationships. Proc. Amer. Soc. Civil Engrs. 95 (HY1) 6537, p 311-327.

Carte, A.E. 1979. Sustained storms on the Transvaal Highveld. S.A. Geogr. Journal, 61(1) p. 39-56.

Changnon, S.A. and Vogel, J.L., 1981. Hydroclimatological character-istics of isolated severe rainstorms. Water Resources Research 17(6) p 1694-1700.

Dixon, M.J., 1977. Proposed Mathematical Model for the Estimation of Areal Properties of High Density Short Duration Storms. Dept. Water Affairs, Tech. Rept. TR78, Pretoria.

Eagleson, P.S., 1978. Climate, soil and vegetation. 2. The distribution of annual precipitation derived from observed storm sequences. Water Resources Research 14(5) p 713-721.

Eagleson, P.S., 1984. The distribution of catchment coverage by stationary rainstorms. Water Resources Research, 20(5) p 581-590.

Huff, F.A., 1967. Time distribution of rainfall in heavy storms. Water Resources Research, 3(14) p 1007-1019.

Huggins, L.F. and Burney, J.R., 1982. Surface runoff, storage and routing. In Hydrologic Modelling of Small Watersheds. Ed. Haan, C.T., Johnson, H.P. and Brakensiek, D.L., Amer. Soc. Agric. Engrs. Mono-graph No.5.

James, W. and Scheckenberger, R., 1983. Storm dynamics model for urban runoff. Intl. Symp. Urban Hydrology, Hydraulics and Sediment control, Lexington, Kentucky. p 11-18.

Keifer, C.J. and Chu, H.H. 1957. Synthetic storm patterns for drainage design. Proc. Amer. Soc. Civil Engrs. 83 (HY4) p 1332-1352,

Mader, G.N., 1979. Numerical study of storms in the Transvaal. S.A. Geogr. Journal, 61(2) p 85-98.

Natural Environment Research Council, 1975. Flood Studies Report, Vol. 1. Hydrological Studies, London, 5 volumes.

Scheckenberger, R., 1984. Dynamic spatially variable rainfall models for stormwater management. M. Eng. Report, McMaster University, Hamilton.

Stephenson, D., 1984. Kinematic study of effects of storm dynamics on runoff hydrographs. Water S.A. October, Vol. 10, No. 4, pp 189-196.

Van Wyk, W. and Midgley, D.C., 1966. Storm studies in S.A. - Small area, high intensity rainfall. The Civil Eng. in S.A., June, Vol. 8 No.6, p 188-197.

Waymire, E. and Gupta, V.L. 1981. The mathematical structure of rainfall representations 3, Some applications of the point process theory to rainfall processes. Water Resources Research, 17(5), p 1287-1294.

Wilson, C.B., Valdes, J.B. and Rodrigues, I.I., 1979. On the influence of the spatial distribution of rainfall in storm runoff. Water Resources Research, 15(2), p 321-328.

Woolhiser, D.A., 1977. Unsteady free surface flow problems. In Mathematical Models for Surface Water Hydrology. Ed. by Ciriani, T.A. Maione, U. and Wallis, J.R., John Wiley & Sons, 423 pp.

CHAPTER 8

CONDUIT FLOW

KINEMATIC EQUATIONS FOR NON-RECTANGULAR SECTIONS

The analysis of flow in conduits is more complicated than for overland flow on account of side friction. Non-rectangular cross sections e.g. trapezoids and circular drains are more difficult than rectangular sections to analyze. Surface width and hydraulic radius become a function of water depth. The sides of the channel (and top in the case of closed conduits) increase friction drag. As far as the form of the basic kinematic equations is concerned the mathematical expressions become more complicated, and numerical solutions are necessary in the majority of cases.

The continuity equation remains

$$\frac{\partial A}{\partial t} + \frac{\partial Q}{\partial x} = q_i \qquad (8.1)$$

or expanding the second term,

$$B\frac{\partial y}{\partial t} + A\frac{\partial v}{\partial x} + v\frac{\partial A}{\partial x} = q_i \qquad (8.2)$$

where the first term is the rate of rise, the second prism storage and the third wedge storage.

The dynamic equation reduces to

$$Q = \alpha AR^M \qquad (8.3)$$

where Q is the discharge rate, α is a function of conduit roughness, q is inflow per unit length, B is the surface width, A is the cross sectional area of flow and R is the hydraulic radius A/P where P is the wetted perimeter. Employing Manning's friction equation,

$$\alpha = K_1 S^{1/2}/n \text{ and } M = 2/3 \qquad (8.4)$$

where K_1 = 1(S.I. units) and 1.486 (ft-sec units)

$\quad\quad$ n = Manning's roughness coefficient

Owing to the greater depths in conduits in comparison with overland flow, lower values of n are applicable. The above equations can be solved for special cases of non rectangular conduits as indicated below.

PART-FULL CIRCULAR PIPES

The cross sectional area of flow in a circular conduit (Fig. 8.1) running part full (Stephenson, 1981) is

$$A = \frac{D^2}{4} \left(\frac{\Theta}{2} - \cos\frac{\Theta}{2}\sin\frac{\Theta}{2} \right) \tag{8.5}$$

and
$$P = D\frac{\Theta}{2} \tag{8.6}$$

Thus if one takes Θ as the variable, the continuity equation becomes

$$\frac{\partial A}{\partial \Theta}\frac{\partial \Theta}{\partial t} + \frac{\partial Q}{\partial x} = q \; ;$$

and

$$\frac{D^2}{8} \left(1 + \sin^2\frac{\Theta}{2} - \cos^2\frac{\Theta}{2} \right) \frac{\partial \Theta}{\partial t} + \frac{\partial Q}{\partial x} = q \tag{8.7}$$

In finite difference form, solving for Θ after a time interval Δt,

$$\Theta_2 = \Theta_1 + \left(q - \frac{\Delta Q}{\Delta x} \right) \frac{8\,\Delta t}{D^2 \left(1 + \sin^2\frac{\Theta}{2} - \cos^2\frac{\Theta}{2} \right)} \tag{8.8}$$

and in terms of the new Θ , since $Q = \alpha AR^{2\,3}$

$$Q = \alpha\frac{D^2}{4} \left(\frac{\Theta}{2} - \cos\frac{\Theta}{2}\sin\frac{\Theta}{2} \right) \left\{ \frac{D}{4}\left(1 - \frac{\cos\frac{\Theta}{2}\sin\frac{\Theta}{2}}{\Theta/2} \right) \right\}^{2\,3} \tag{8.9}$$

In order to simulate flow and depth variations in pipes, the latter two equations are applied at successive points for successive time intervals.

In addition to analysis of flows in pipes, the methods can be applied to design by successive analysis. When designing storm drain collection systems there are many approaches (Yen and Sevuk, 1975). It is in normal practice not necessary to consider surcharged conditions in a design unless a dual system (major and minor conduits) is employed. If pipes are designed to run just full at their design capacity, then they will run part full for any other design storm duration. The higher up the leg a pipe length is, the shorter will be the concentration time, or time to flow equilibrium. The design storm duration will equal the concentration time of the drains down to the pipe in question. Any subsequent pipes will have larger concentration times and consequently a lower storm intensity.

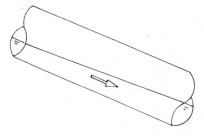

Fig. 8.1   Cross section through part-full pipe

COMPUTER PROGRAM FOR DESIGN OF STORM DRAIN NETWORK

The preceding scheme was employed in a program for analysing the flow in each pipe in a drainage network the plan of which is specified by the designer. The engineer must pre-select the layout, subdivision of catchment, position of inlets and grades. The grades will in general conform to the slope of the ground.

It is necessary to simulate overland flow and each upper drain in order to size any lower drain. Such analysis can only be done practically by digital computer using numerical solutions of the flow equations. Many calculations are necessary for complex networks. A limitation on the maximum time interval for numerical stability implies many iterations until equilibrium flow conditions are reached for each pipe design. In addition, a number of different storm durations must be investigated for each pipe. A simple and efficient iterative procedure was therefore sought in order to minimize computer time. The kinematic form of the flow equation was employed to ensure this. The emphasis throughout the program is simplicity of data input and minimization of computational effort. Some accuracy is sacrificed by the simplifications but the overriding assumption of precipitation pattern is probably more important.

The design method (Stephenson, 1980) proceeds for successive pipes, the diameters of which are calculated previously. It is assumed the network layout is specified, and the pipe grades are dictated by the ground slope. Starting at the top ends of a drainage system, the program sizes successively lower pipes. Thereby each pipe upstream of the one to be designed is pre-defined. It is necessary to investigate storms of different duration and corresponding intensity of flow to determine the design storm resulting in maximum flow for the next pipe.

It is assumed that the design storm recurrence interval is preselected. The intensity-duration relationship is then assumed to be of the form

$$i_e = \frac{a}{b + t_d} \tag{8.10}$$

By selecting storms of varying duration t_d, and simulating the flow buildup down the drains, the program can select a storm which will result in the maximum peak flow from the lower end of the system. That discharge is the one to use for sizing the next lower pipe. Thus the program proceeds from pipe to pipe until the entire network is designed.

The program is limited in application to selection of drain pipe diameters for a simple gravity collecting system, and uses kinematic theory and the limitations of the theory should be recalled. It should be noted that for major pipes it may become necessary to allow for backwater and routing effects (Barnes, 1967). The program does not optimize the layout (Argamon et al. 1973; Merritt and Bogan, 1973). Nor is surcharge (Martin and King, 1981) or detention storage considered here.

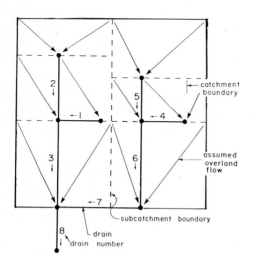

Fig. 8.2   Layout plan of drainage network sized in example

Program description

Pipes are assumed to flow initially at a depth corresponding to a subtended angle of 0.2 radians at the centre. The corresponding flow is very low, but this assumption avoids an anomaly for the case of zero depth when the numerical solution of the explicit equation is impossible.

Inflow from subcatchments is assumed to occur along the full length of the respective pipe, i.e. subcatchment breadth is assumed to be equal to pipe length. This affects overland flow time to some extent. If necessary (if flow is sensitive to storm duration) the subcatchment friction factor could be adjusted to give the correct overland flow time.

The computer program, written in FORTRAN for use in conversational mode on a terminal connected to an IBM 370 machine, is appended. The input format is described below. Data is read in free format and can be input on a terminal as the program stands.

First line of data:

M, A, B, E, IN, IR, II, G.

Second and subsequent lines of data (one line for each length of pipe):

X(I), S(I), Z(I), C(I), SO(I), EO(I), IB(I).

The input symbols are explained below:

M – The number of pipes: the number of pipes should be minimized for computational cost minimization. For computational accuracy the pipes should be divided into lengths of the same order of magnitude. It is convenient to make the pipe lengths equal to the distance between inlets. Inlets between 10 and 200 m apart are normally sufficient for computational accuracy. There should be at least two pipes in the system.

A,B – Precipitation rate i is calculated from an equation of the form $i = A/(B + t_d)$ where t_d is the storm duration and B is a regional constant (both in seconds). A is a function of storm return period and catchment location and its units are in m if SI units are used, and ft if ft–lb–sec units are used.

E – Pipe roughness. This is analogous to the Nikuradse roughness and E is measured in m or ft. It is assumed in the program that all pipes have the same roughness. A conservative figure of at least 0.001 m (0.003 ft) is suggested to account for surface deterioration with time due to erosion, corrosion or deposits.

IN, – For each pipe sizing computation various storm durations
IR are investigated, ranging from IU1 to IU2 in steps of IR (all in seconds). The smallest storm duration IU1 is set equal to the overland flow time for an upper pipe of the previous pipe design storm duration for subsequent pipes down a leg. The number of storm durations investigated is specified by IN and the increment in trial storm duration is specified by IR. Thus IU2 = IU1 + IN*II. The accuracy of the computations is affected by the number of trial storm durations. A value

of IN between 3 and 10 is usually satisfactory. The upper limit can be estimated beforehand from experience or by trial (if all design storm durations turn out to be less than the IU2 specified then the IN selected is satisfactory).

II — The computational time and cost is affected by the time increment of computations II (seconds). The maximum possible value is dependent on the numerical stability of the computations. A value equal to the minimum value of

$$(\frac{B}{C(I)A})^{0.4} \ (\frac{Z(I)E}{100 \ X \ (1) \ \sqrt{(S(I)G)}})^{0.6}$$

will normally be satisfactory (of the order of 60 to 300 seconds).

G — Gravitational acceleration (9.8 in SI units and 32.2 in ft-sec units).

The pipe data are next read in line by line for M pipes. As the program stands, 98 individual pipes are permitted, and any number of legs subject to the maximum number of pipes.

X(I) The pipe length in m or ft, whichever units are used. An upper limit on individual pipes of 200m is suggested for computational accuracy and a lower limit of 10m for optimizing computer time.

S(I) The slope of the pipe in m per m or ft per ft.

Z(I) The surface area contributing runoff to the pipe in m or ft

C(I) The proportion of precipitation which runs off (analogous to the 'C' in the Rational formula).

SO(I) The overland slope of the contributing area, towards the inlet at the head of the pipe.

EO(I) The equivalent roughness of the overland area in m or ft depending on units employed.

IB(I) The number of the pipe which is a branch into the head of pipe I.

For no branch, put IB(I) = 0

For a header pipe at the top of a leg, put IB(I) = -1.
Only one branch pipe per inlet is permitted.
More must be accommodated by inserting short dummy pipes between.
The order in which pipes are tabulated should be obtained as follows:

Computer Program for Storm Network Pipe Sizing

```
L.0001          DIMENSION P(99),Q(99),S(99),X(99),D(99),AT(99),Z(99),QQ(99),UU(99),C(99)
L.0002          DIMENSION EO(99),SO(99),QO(99),IB(99),IA(99),QP(99)
L.0003          READ(9,5)M,A,B,E,IN,IR,II,G
L.0004    5     FORMAT(8Y)
L.0005   10     FORMAT(7Y)
L.0006          DO 15 I=1,M
L.0007          READ(9,10)X(I),S(I),Z(I),C(I),SO(I),EO(I),IB(I)
L.0008          IF(IB(I).GT.1)GO TO 13
L.0009          IF(IB(I).LT.0)GO TO 12
L.0010          IB(I)=99
L.0011          GO TO 13
L.0012   12     IB(I)=99
L.0013          IA(I)=99
L.0014          GO TO 15
L.0015   13     IA(I)=I-1
L.0016   15     CONTINUE
L.0017          C(99)=0.
L.0018          QP(99)=0.
L.0019          QP(1)=0.
L.0020          M1=M-1
L.0021          DO 25 I=1,M
L.0022          IF(IA(I).LT.99)GO TO 25
L.0023          PO=C(I)*A/(B+500)
L.0024          DO 20 J=1,10
L.0025   20     PO=C(I)*A/(B+(Z(I)/X(I)*EO(I)**.167/7.7/(SO(I)*G)**.5)**.6/PO**.4)
L.0026          IUI=C(I)*A/PO-B
L.0027          UU(I)=IUI
L.0028          PP1=PO*Z(I)
L.0029          QQ(I)=PP1
L.0030          D(I)=(PP1*E**.167/7.7/(S(I)*G)**.5/3.141*4**1.667)**.375
L.0031   25     CONTINUE
L.0032          DO 300 M2=1,M1
L.0033          IU1=IUI
L.0034          IF(IA(M2+1).GE.99)GO TO 300
L.0035          QQ(M2+1)=0.
L.0036          IU2=IU1+IN*IR
L.0037          DO 200 IU=IU1,IU2,IR
L.0038          DO 30 I=1,M
L.0039          BIU=B+IU
L.0040          QO(I)=Z(I)/10./X(I)*C(I)*A/(BIU)
L.0041          AT(I)=.2*3.141
L.0042   30     Q(I)=0.
L.0043          DO 120 IT=1,IU,II
L.0044          M3=M2+1
L.0045          DO 32 I=1,M3
L.0046          QOV=(7.7*(SO(I)*G)**.5/EO(I)**.167)**.6*1.667*II*QO(I)**.4
L.0047          QO(I)=QO(I)+QOV*(C(I)*A/(B+IU)-QO(I)/Z(I)*X(I))
L.0048   32     P(I)=QO(I)*X(I)
L.0049          DO 100 I=1,M2
L.0050   35     IF(IT.LE.1)GO TO 50
L.0051   40     A1=(P(I)-Q(I)+QP(IA(I))+QP(IB(I)))/X(I)*II*8./D(I)/D(I)
L.0052   45     AT(I)=A1/(1-(COS(AT(I)/2.))**2.+(SIN(AT(I)/2.))**2.)+AT(I)
L.0053   50     A2=AT(I)/2.
L.0054          C1=7.7*SQRT(G*S(I))/E**(1./6.)
L.0055          C2=Q1*D(I)**2./4.*(A2-COS(A2)*SIN(A2))
L.0056  100     Q(I)=Q2*(D(I)/4.*(1.-COS(A2)*SIN(A2)/A2))**(2./3.)
L.0057          DO 110 I=1,M2
L.0058  110     QP(I)=Q(I)
L.0059  120     CONTINUE
L.0060          Q(M2+1)=Q(IB(M2+1))+Q(M2)+P(M2+1)
L.0061          IF(Q(M2+1).LE.QQ(M2+1))GO TO 200
L.0062          QQ(M2+1)=Q(M2+1)
L.0063          UU(M2+1)=IU
L.0064  200     D(M2+1)=(QQ(M2+1)*E**(1./6.)/7.7/SQRT(S(M2+1)*G)/3.141*4.**(5./3.))**(3./8.)
L.0065          IF(IA(M2+1).LT.99)GO TO 290
L.0066          IU1=IUI
L.0067          GO TO 300
L.0068  290     IU1=UU(M2+1)
L.0069  300     CONTINUE
L.0070          WRITE(5,350)
L.0071  350      FORMAT(' STORM SEWER DESIGN')
L.0072          WRITE(5,60)
L.0073   60     FORMAT(' PIPE LENGTH  DIA  GRADE  DSFLO/S  STORM S    AREA')
L.0074          DO 400 I=1,M
L.0075  400     WRITE(5,70)I,X(I),D(I),S(I),QQ(I),UU(I),Z(I)
L.0076   70     FORMAT(I6,F7.0,F6.3,F6.4,F9.3,F8.0,F9.0,F9.0)
L.0077          WRITE(5,80)M,A,B,E,IU2,IR,II
L.0078   80     FORMAT(' DATA'I6,F5.3,F5.0,F5.4,3F6.0)
L.0079          STOP
L.0080          END
```

                      COMPUTER OUTPUT FOR SAMPLE RUN

```
L.0001    STORM SEWER DESIGN
L.0002    PIPE LENGTH  DIA  GRADE  DSFLO/S  STORM S    AREA
L.0003      1   100.  .576 .0020    .244    1014.   20000.
L.0004      2   150.  .514 .0040    .255     911.   20000.
L.0005      3   200.  .643 .0040    .462    2068.   40000.
L.0006      4   100.  .415 .0020    .102     772.   10000.
L.0007      5   100.  .574 .0040    .342    2068.   40000.
L.0008      6   200.  .619 .0040    .417    2068.   10000.
L.0009      7   200.  .853 .0020    .696    2068.   40000.
L.0010      8   100.  .905 .0050   1.287    2068.   20000.
L.0011    DATA   8 .0751440..0010  2963    300      60
```

After drawing out a plan of the catchment with each pipe, the longest leg possible is marked, starting from the outfall, then success-ively shorter legs on first the longest, then successively shorter pipes. Now the pipes are numbered in the reverse over, starting at the top of the shortest leg etc. Proceed down each leg with the numbering until a junction is reached. Never proceed past a branch which has not been tabulated previously. In this way all pipes leading into a pipe will have had their diameters calculated before the next lower pipe is designed.

Sample Input

The data are in metres and are taken from Fig. 8.2

8	.075	1440	.001	3	300	60	9.8
100	.002	20000	.4	.005	.01	−1	
150	.004	20000	.4	.003	.01	−1	
200	.004	40000	.4	.003	.01	1	
100	.002	10000	.3	.005	.02	−1	
100	.004	40000	.4	.003	.01	−1	
200	.004	10000	.5	.005	.01	4	
200	.002	40000	.4	.002	.01	0	
100	.005	20000	.4	.003	.01	3	

TRAPEZOIDAL CHANNELS

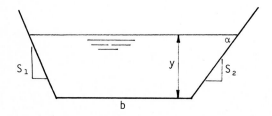

Fig. 8.3   Trapezoidal channel geometry

For trapezoidal channels the hydraulic equations become

$A = y (b + y/S_1 + y/S_2)$

$P = b + y [\sqrt{(1 + 1/S_1^2)} + \sqrt{(1 + 1/S_2^2)}]$

In particular for a vertical sided rectangular channel of limited width b, employing the Manning equation.

$A = yb,$

$P = b + 2y$

$Q = \alpha yb \left(\dfrac{yb}{b+2y}\right)^{2/3}$

$= \alpha (yb)^{5/2} / (b+2y)^{2/3}$

The analysis of flow in channels must generally be done numerically. The channel is divided into reaches and a suitable time step selected to simulate flow and depth variations. The continuity and friction equations are applied conjunctively to calculate increase in water depth and flow rate respectively. The method can be employed for catchment channel flow simulations. Many natural channels can be approximated by a trapezoid, or else a number of trapezoids. A channel plus flood plane can be represented by two trapezoids at different bed levels, the flood plane being at the top of the banks of the channel. The roughness, and hydraulic radius, and consequently the velocity will differ from channel to overbank and this can be accounted for.

COMPARISON OF KINEMATIC AND TIME-SHIFT ROUTING IN CONDUITS

Whereas overland flow time lag may be predicted quite differently using kinematic or time lag methods, in the case of conduits, time lag often provides a sufficiently accurate assessment of flow. That is, owing to the confined cross section of a conduit, flow is more inclined to emerge at the same rate that it enters a conduit, and travel time approximates reaction time sufficiently well.

In stormwater drainage, runoff hydrographs from overland flow constitute the essential input to hydraulic conduits; e.g. pipes, channels, culverts etc. The overland flow hydrographs are attenuated further as they travel through the conduits. In a stormwater drainage network, where conduits and manholes are interlinked to carry water from different subcatchments onto a major outlet, hydrograph attenuation through the conduits is very important. Hydrographs from conduits leading to the same manhole have to be summated for designing hydraulic structures or conduits downstream or for studying the behaviour of an existing network under certain conditions. The magnitude of the hydrograph peaks as well as their relative time positions are important for the accurate assessment of design flows.

Various methods exist for routing runoff hydrographs through closed conduits. The most commonly used are time shift methods. A time shift method shifts the entire hydrograph in time without any storage consider-ations for attenuation. The time shift or lag time is calculated by dividing the length of the conduit by the velocity of the water in the conduit. This velocity is usually taken to be the velocity of water in the conduit when the conduit is almost full under steady conditions.

Storage balance methods are also used for routing. They apply mass balance equations across the conduit. Such equations are solved by either explicit or implicit schemes. Both time shift and storage routing methods ignore non-uniform flow and dynamic effects in the system. Other methods for hydrograph routing include routing through conduits using the kinematic equations or even the dynamic equations of flow.

The use of the kinematic equations for routing requires comparatively large computational effort in comparison with time shift as the equations have to be solved at close grid points along the conduit over short time increments. Most existing drainage models use time shift methods and since the solution of the kinematic equation is tedious it may in some cases be unwarranted.

Section Geometry and Equations for Conduits

Two section configurations are studied here, one a circular section and the other a trapezoid. Both sections are assumed to be partly full as dynamic effects of the system are not studied. For the pipe this implies that the depth of flow is always less than the pipe diameter while for the trapezoid its sides are assumed to be high enough to allow any depth of water.

For partially filled closed conduits, i.e. where no lateral inflow exists along the conduit, the kinematic continuity equation is:

$$\frac{\partial q}{\partial x} + \frac{\partial a}{\partial t} = 0 \qquad (8.11)$$

where q is discharge (m^3/s), a is cross sectional area of flow (m^2), x is distance along the conduit from the inlet (m) and t is time (s).

In kinematic theory discharge can be assumed to be a function of flow depth as the friction slope is assumed to equal the bed slope. This enables the use of uniform flow equations expressed in terms of bed slope instead of friction slope. Such equations are usually described in the following form:

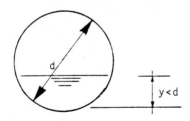

(a)   Pipe

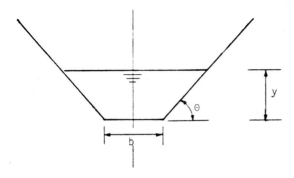

(b)   Trapezoid

Fig. 8.4   Conduit Sections

$$q = \alpha \, a \, R^{m-1} \tag{8.12}$$

where α and m   are friction flow coefficients depending on the uniform flow equation used, R is the hydraulic radius of the section, i.e. a/p (m) and p is the wetted perimeter of the section (m).

$$\alpha = \frac{1}{n} \, S^{1/2} \text{and } m = 5/3 \tag{8.13}$$

where n = Manning's roughness coefficient and S = bed slope.

Inserting the values of α and m from   8.13   in equation 8.12 yields:

$$q = \frac{1}{n} \, S^{\frac{1}{2}} \, \frac{a^{5/3}}{p^{2/3}} \tag{8.14}$$

The geometry of the conduits is described by equations 8.15 − 8.18.

$$A = \frac{d^2}{4} \cos^{-1} \left[\frac{d-2y}{d} \right] - \left[\frac{d}{2} - y \right] \left[dy - y^2 \right]^{\frac{1}{2}} \qquad (8.15)$$

Pipe

$$P = d \cos^{-1} \left[\frac{d-2y}{d} \right] \qquad (8.16)$$

$$A = by + y^2 \tan (90 - \Theta) \qquad (8.17)$$

Trapezoid

$$P = b + 2y \sec (90 - \Theta) \qquad (8.18)$$

The equations 8.11 and 8.14 were reduced to a dimensionless form by Constantinides (1983) with the choice of suitable variables. The dimensionless equations are then solved for different conduit sections and input hydrographs. The kinematic equations are solved in their dimensionless form to facilitate generalization of results in terms of constant parameters that are functions of the input parameters. The use of the dimensionless equations reduces computational effort as the number of cases to study reduces greatly.

The variables q, a, x and t are reduced to the dimensionless variables Q, A, X and T by dividing them by appropriate variables with identical units as follows:

For the pipe,

$$Q = q/q_m \qquad (8.19)$$

$$A = a/d^2 \qquad (8.20)$$

$$P = p/d \qquad (8.21)$$

$$Y = y/d \qquad (8.22)$$

For the trapezoid,

$$Q = q/q_c \qquad (8.23)$$

$$A = a/b^2 \qquad (8.24)$$

$$P = p/b \qquad (8.25)$$

$$Y = y/b \qquad (8.26)$$

and for both sections

$$X = x/L \qquad (8.27)$$

$$T = t/t_k \qquad (8.28)$$

where q_m is the maximum flow capacity of the pipe (m^3/s), $\frac{0.33528}{n} S^{1/2} d^{8/3}$ q_c is a discharge variable, being a function of friction coefficients Θ, m and bottom width of trapezoid, b (m^3/s) i.e. $q_c = \frac{1}{n} S^{1/2} b^{8/3}$, t_k is a time constant (s) and L is the length of the conduit (m).

To define the discharge and time constants appropriately the dimensionless kinematic equations are obtained by substituting the dimensionless variables in the continuity equation i.e. for the pipe,

$$\frac{\partial (Q \, q_m)}{\partial (XL)} + \frac{\partial (Ad^2)}{\partial (Tt_k)} = 0 \qquad (8.29)$$

Rearranging yields:

$$\frac{q_m \cdot t_k}{Ld^2} \quad \frac{\partial Q}{\partial X} + \frac{\partial A}{\partial T} = 0 \tag{8.30}$$

Furthermore by defining the time constant as in equation 8.31 reduces equation 8.30 to the dimensionless equation 8.33. Similarly for the trapezoid the time constant is defined in equation 8.32.

For the pipe:

$$t_k = \frac{Ld^2}{q_m} \tag{8.31}$$

For the trapezoid:

$$t_k = \frac{Lb^2}{q_c} \tag{8.32}$$

where the dimensional continuity equation is:

$$\frac{\partial Q}{\partial X} + \frac{\partial A}{\partial T} = 0 \tag{8.33}$$

Similarly, the uniform flow equation 8.14 can be reduced to its dimensionless form, i.e.

for the pipe:

$$Q \, q_m = \frac{1}{n} \, S^{\frac{1}{2}} \, \frac{(Ad^2)^{5/3}}{(Pd)^{2/3}} \tag{8.34}$$

where the maximum carrying capacity of a pipe can be shown to be

$$q_m = 0.335282 \, \frac{1}{n} \, S^{\frac{1}{2}} \, d^{8/3} \tag{8.35}$$

Substituting in equation 8.34 and rearranging yields:

$$Q = \frac{1}{0.335282} \, \frac{A^{5/3}}{P^{2/3}} \tag{8.36}$$

For the trapezoid the uniform flow equation reduces to:

$$Q \, q_c = \frac{1}{n} \, S^{\frac{1}{2}} \, \frac{(Ab^2)^{5/3}}{(Pd)^{2/3}} \tag{8.37}$$

Defining q_c as in equation 8.38 reduced equation 8.37 to the dimensionless flow equation for the trapezoid given in equation 8.39:

$$q_c = \frac{1}{n} \, S^{\frac{1}{2}} \, b^{8/3} \tag{8.38}$$

For the trapezoid:

$$Q = \frac{A^{5/3}}{P^{2/3}} \tag{8.39}$$

Equations for t_k for both sections can be obtained by substituting equation 8.35 and 8.38 into equation 8.31 and 8.32. Similarly, for obtaining the dimensionless area and perimeter variables (A and P) equations 8.20, 8.21 and 8.22 are substituted in equations 8.15 to 8.18. The resulting expressions are summarised below.

Pipe

$$A = \frac{1}{4} \cos^{-1} (1-2Y) - (\frac{1}{2} - Y) \cdot (Y - Y^2)^{1/2} \qquad (8.40)$$

$$P = \cos^{-1} (1-2Y) \qquad (8.41)$$

$$t_k = \frac{L}{0.335282\frac{1}{n} S^{\frac{1}{2}} d^{2/3}} \qquad (8.42)$$

Channel

$$A = Y + Y^2 \tan (90-\theta) \qquad (8.43)$$

$$P = 1 + 2Y \sec (90-\theta) \qquad (8.44)$$

$$t_k = \frac{L}{\frac{1}{n} S^{\frac{1}{2}} b^{2/3}} \qquad (8.45)$$

Two shapes of inflow hydrographs are routed through the conduits, one a uniform and the other a triangular time distribution. These two time distributions were chosen as they represent extreme cases, i.e. a natural runoff hydrograph, from overland flow, would have a shape between these two extremes depending on the rainfall and catchment characteristics.

In addition to the shapes the hydrographs were assumed to have a variety of durations and intensities. Fig. 8.5 illustrates the inflow hydrographs in their dimensionless form.

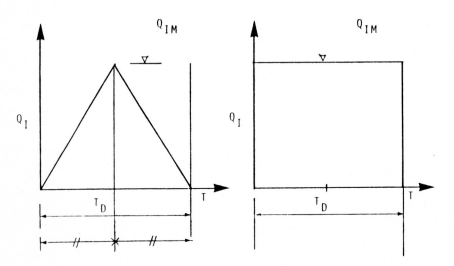

Q_{IM} is the maximum discharge factor or inflow factor and $T_D = t_d/t_k$, where t_d is duration.

Fig. 8.5 Different dimensionless inflow hydrographs

The dimensionless equation for speed of propagation is

$$\frac{dX}{dT} = \frac{1}{C_1} \left(\frac{A}{P}\right)^{2/3} \left(\frac{A}{P} \frac{5}{3} - \frac{2}{3} \frac{A}{P} \frac{1}{\partial A/\partial P}\right) \tag{8.46}$$

where C_1 = 0.335282 for the pipe
        C_1 = 1.0 for the trapezoid

where for pipes:

$$\frac{\partial A}{\partial P} = 2(Y - Y^2) \tag{8.47}$$

and for trapezoids:

$$\frac{\partial A}{\partial P} = \frac{1}{2} \cos (90-\theta) + Y \sin (90-\theta) \tag{8.48}$$

Equation 8.46 is a function of the depth coefficient, Y. it was solved in terms of Y using a computer model.

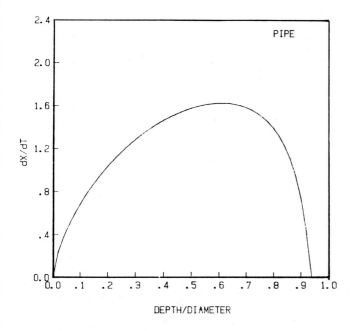

Fig. 8.6   Dimensionless propagation speed of a disturbance in partially filled pipes

160

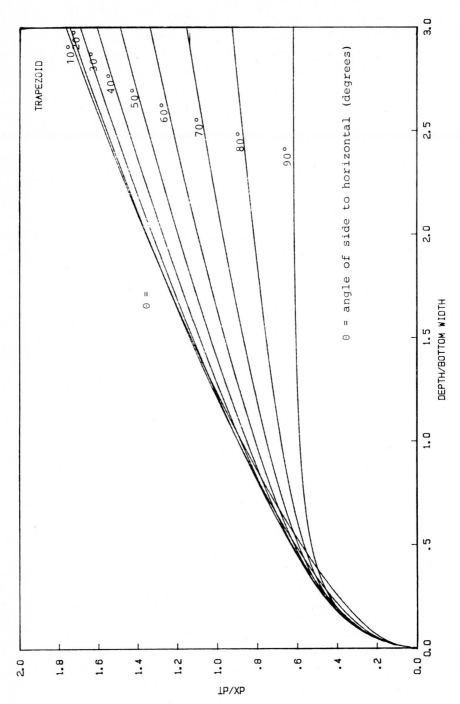

Fig. 8.7   Dimensionless propagation speed of a disturbance in trapezoids

Typical results are given in Fig. 8.6 and 8.7. It can been seen from Fig. 8.6 the maximum dimensionless propagation speed in a pipe is 1.63 and occurs when the depth over diameter ratio is 0.62. Fig. 8.7 shows that for the trapezoid the dimensionless propagation speed increases with an increase in the depth over bottom width ratio. It is necessary to know for both the pipe and the trapezoid the maximum depth over diameter and depth over bottom width ratios respectively in order to assess the maximum propagation speed, $(dX/dT)_m$, during any single simulation. It should be noted that for the pipe any simulation, where the depth over diameter ratio exceeds 0.62, will have a $(dX/dT)_m$ of 1.63.

The maximum depth of flow in the conduit to be encountered during simulation will be a function of the maximum inflow discharge at the inlet, as the hydrograph will attenuate as it travels away from the inlet. The maximum dimensionless depth of flow in the conduit (Y) is related to the maximum dimensionless inflow discharge, or inflow factor (Q_{IM}), by equations 8.36 and 8.39.

Equation 8.46 yields maximum propagation speeds for different inflow factors.

Computer Simulation

A computer model was developed for solving the dimensionless kinematic equations for closed conduits. The model routes dimensionless inflow hydrographs through the conduits to produce dimensionless outflow hydrographs at the outlet. The dimensionless hydrographs were then studied to evaluate the effects that a section of fixed geometry and length has in attentuating inflow hydrographs of varying discharge and duration.

For every inflow factor and inflow hydrograph distribution different dimensionless storm durations were assumed. The dimensionless storm durations were assumed. The dimensionless storm durations are defined as the storm duration over the time constant ratio, i.e.

$$T_D = \frac{t_d}{t_k} \tag{8.49}$$

Values of T_D varied from 0.2 to 10 according to the inflow time distribution and section type. The following observations are made:
a) Simulations indicated the lag time of the outflow relative to the inflow hydrograph decreases with an increase of inflow hydrograph duration (for a constant inflow factor). The reason for this is that longer duration inflows imply higher inflow volumes. Hydrographs with lower volumes tend to spread more within the conduit resulting in lower water depths which in turn result in lower flow velocities and propaga-

tion of disturbance speeds. This inevitably increases their lag time. The same argument explains the second observation, i.e.

b) The ratio of peak at the outlet over peak at the inlet increases with increasing storm duration (for a constant inflow factor) or in other words inflow hydrographs of smaller storm duration undergo higher discharge attenuation than hydrographs of longer duration. The reason for this is the same as in a), i.e. lower volumes spread more than bigger volumes resulting in lower depths of flow and thus lower discharges.

c) The lag time for an inflow hydrograph of fixed duration decreases with higher inflow factors. The reason for this is identical to a) as higher inflow factors imply higher volumes of water.

d) Peak flow attenuation is higher for small inflow factors (for a constant inflow duration) than for high inflow factors, the reason being the same as for observation b).

Further deductions from the results can be made by representing the printout results in the form of graphs. This is done in subsequent sections.

Criteria for choosing between Time Shift and Kinematic Routing

One of the main objectives of this study was to develop a method for assessing whether time shift methods can be used without having to resort to routing methods. The main assumption behind time shift methods is the preservation of the hydrograph ordinates without any attenuation. To accept time shift methods, therefore, the hydrograph attenuation that would happen in a real life situation must be small. One must therefore decide what are acceptable limits of attenuation. As hydrograph attenuation differs throughout the hydrograph duration one usually refers to peak attenuation. In this study a peak attenuation of 10% is taken to be the maximum peak attenuation that can be ignored. This value, although arbitrarily defined, is based on the fact that more accurate determination of runoff is not justified due to the corresponding inaccuracies in input determination. Furthermore, in a drainage system consisting of various conduits interlinked in a network, tolerating a bigger peak attenuation can result in a gross overestimation of the outflow peak. This occurs since a small peak attenuation is propagated downstream through various conduits and doing that increases in magnitude.

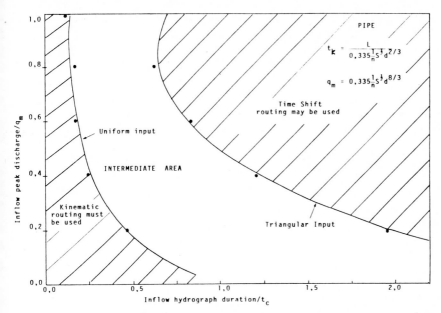

Fig. 8.8   Diagram indicating when time shift routing can be used with partially filled pipes.

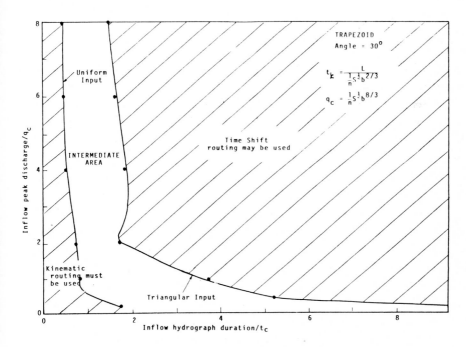

Fig. 8.9   Diagram indicating when time shift routing can be used with trapezoids at angle of side to horizontal of 30°

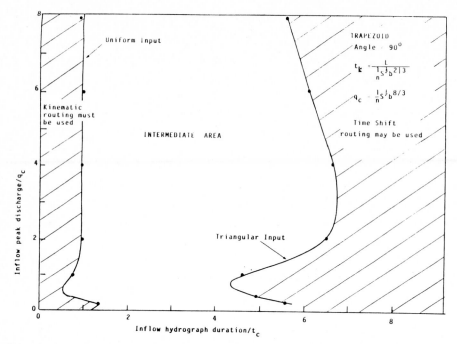

Fig. 8.10   Diagram indicating when time shift routing can be used with trapezoids at angle of side to horizontal of 90°

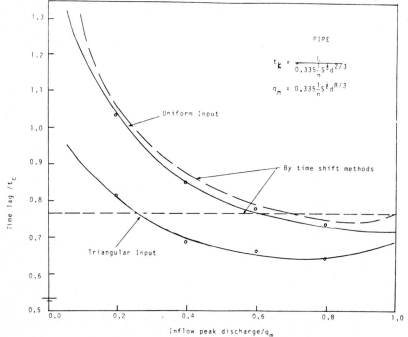

Fig. 8.11   Time lag for hydrographs routed through partially filled pipes

Having decided on an acceptable peak attenuation to be neglected it is assumed that kinematic routing describes accurately routing in a real life situation. The results obtained by kinematic routing are employed to assess the conditions under which time shift methods are acceptable, i.e. in this case the conditions under which the peak attenuation is lower than 10%. To do this the results were used to obtain a dimensionless inflow duration for a 0.9 outflow to inflow peak ratio for every type of section and inflow factor. The dimensionless inflow duration was obtained either by linear interpolation or whenever thought necessary by plotting dimensionless duration against outflow to inflow peak ratio and obtaining the dimensionless duration for a peak ratio of 0.9, the peak ratio of 0.9 corresponding to a 10% peak attenuation. The results are summarised in Figs. 8.8 – 8.10.

Lag Time for Routing Hydrographs Using Time Shift Methods

Using a similar method to the above the dimensionless time lag of hydrographs with a peak attenuation of 10% was obtained. The dimensionless lag times are summarised in Figs. 8.11 to 8.13 for pipes and selected trapezoids.

A dotted line represents lag times as obtained by time shift methods for comparison purposes.

Comparison of Methods for Evaluating Lag Time

Two assumptions are currently popular for calculating the time lag of a hydrograph to be routed by time shift methods. The time lag is either assumed to be the length of the conduit divided by the velocity of the water when the conduit is discharging at full capacity or it is assumed that the time lag is the length of the conduit divided by the velocity of water in the conduit corresponding to the maximum discharge of the inflow hydrograph.

Method 1   ($T_{Lp} = L/(q_m/a_m)$.) The time constant (t_k) for the pipe is given by equation 8.31. The dimensionless time lag is thus

$$\frac{t_{LP}}{t_k} = \frac{a_m}{d^2}$$

where a_m/d^2 is the dimensionless flow area for a pipe discharging at maximum capacity. Substituting for a_m/d^2

$$\frac{t_{LP}}{t_k} = 0.7653 \tag{8.50}$$

This gives the dimensionless lag time for a pipe and plots in Fig. 8.11 as a straight line. As can be seen the lag time calculated by this method over-estimates the true value for high inflow factor values (bigger than 0.6) and grossly underestimates it for low inflow factor values (lower than 0.25). For intermediate inflow factor values this method yields time lags which lie between the range set up by the two different input distributions.

Method 2 $(t_{LP} = L/(q_{im}/q_m).)$ The following relationship holds for the dimensionless time lag:

$$\frac{t_{LP}}{t_k} = \frac{A_i}{(q_{im}/q_m)} \tag{8.51}$$

where A_i is the dimensionless flow area corresponding to the maximum discharge of the inflow hydrograph (q_{im}).

Equation 8.51 was solved in the following way to express (t_{LP}/t_k) in terms of the inflow factor (q_{im}/q_m). The dimensionless water depth (Y_i) corresponding to the flow depth A_i is solved knowing the inflow factor and a Newton-Raphson iterative scheme, using equation 8.36.

Y_i is used to solve for A_i and consequently for t_{LP}/t_k. The calculated values of t_{LP}/t_k are plotted in Fig. 8.11 for comparison. It can be seen that this present method yields time lags closely resembling the results obtained from kinematic theory using uniform input hydrographs. This occurs as uniform input hydrographs (which do not attenuate significantly — 10% only) maintain an approximately constant depth through their travel through the conduit, thus having a speed of flow similar to that calculated by the existing method. The fact that time lag as developed by kinematic routing is slightly less than that using the present method is because some attenuation (10%) occurs during routing for producing the results.

Time Lag for Trapezoids

Method (1) as outlined above is not applicable for trapezoids in this study as they are assumed to be deep enough to accommodate incoming hydrographs of any discharge. As their depth is not restricted one cannot talk of maximum discharge through trapezoids. Method (2), however, can be used to express the dimensionless lag time (t_{LP}/t_k) in terms of the inflow factor (q_{im}/q_c) to compare time lags with the present method with the results shown in Figs. 8.12 and 8.13.

The time constant t_k is given by equation 8.32 for the trapezoid. This yields the dimensionsless time lag.

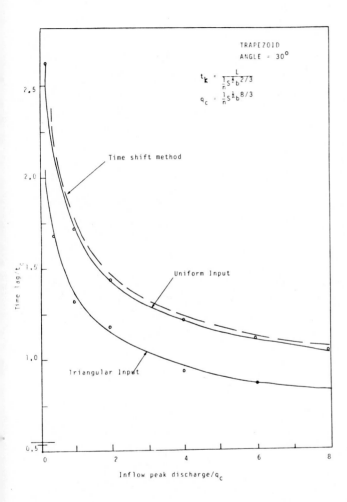

Fig. 8.12   Time lag for hydrographs routed through trapezoids with angle of side to horizontal of 30°

168

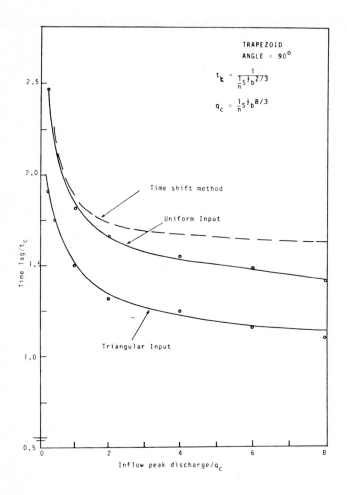

Fig. 8.13 Time lag for hydrographs routed through trapezoids with angle of side to horizontal of 90°

$$\frac{t_{LP}}{t_k} = \frac{A_i}{(q_{im}/q_c)} \qquad (8.52)$$

This equation is solved to yield the ratio t_{LP}/t_k for different values of the inflow factor. Note that the relationship will differ for different angles for the trapezoid as the dimensionless flow area is a function of the side angle. The results are plotted in Figs. 8.12 and 8.13 together with the kinematic routing results for comparison purposes. As can be seen (the dashed lines) the time lags from the present method are slightly higher but closely resemble the ones from kinematic routing using a uniform input. Note that this was also the case for the pipe. The reasons for their resemblance are similar to those for the pipe and are discussed in the previous section.

It can be seen from Figs. 8.8 to 8.10 that the dimensionless inflow duration is much more critical than the dimensionless inflow peak discharge for determining whether time shift methods can be used. This is more apparent in the case of trapezoids where the 10% peak attenuation curves appear almost vertical for dimensionless inflow peak discharge values greater than 2.0.

Furthermore, it can be seen that the dimensionless infow duration decreases with increasing inflow factor. This is expected as inflow hydrographs with a similar inflow factor need bigger durations than ones with a higher inflow factor for both inflow hydrographs to have similar volumes. As was discussed earlier, higher inflow volumes will imply smaller peak attenuation, other parameters being constant, one exception to this observation being the pipe for inflow factors higher than 0.8. It can be seen from Fig. 8.8 that as the inflow factor approaches unity the dimensionless duration (causing a 10% peak attenuation to the inflow hydrograph) increases.

This is probably due to the fact that a pipe discharges more when not flowing full as already discussed.

It will also be noted that for trapezoids and discharge inflow factors of less than 2.0 the dimensionless inflow duration increases sharply as the dimensionless discharge decreases. This is probably due to the fact that at low depths of flow side friction effects cause a stability effect on the flow highly attenuating peak discharges. This in turn implies higher inflow durations for maintaining a peak attenuation of 10%.

For a constant inflow factor, the inflow duration (implying a 10% peak attenuation of the routed hydrograph) is bigger for the triangular distribution than for the uniform one. This is to be expected as a triangular distribution has a lower inflow volume than a uniform one, both distributions having the same duration and inflow factors.

The triangular distribution would therefore need a greater duration (for a constant inflow factor) or a greater inflow factor (for a constant duration) to yield a similar result to the uniform distribution. Note that a constant volume will not imply identical results between the two distributions as the shape also plays an important role in the routing; for example:

From Fig. 8.8, for a pipe and an inflow factor of 0.6, the corresponding dimensionless durations resulting in a 10% peak attenuation of the inflow hydrograph, are found to be 0.18 for a uniform input distribution and 0.82 for a triangular distribution. This implies that the triangular distribution has a bigger inflow volume than the uniform one in the ratio of:

$$\frac{\frac{1}{2}(0.6 \times 0.82)}{0.6 \times 0.18} = 2.28$$

This ratio (triangular to uniform volume) varies depending on the inflow factor and type of section but is always found to be more than unity. This implies further that a uniform time distribution is attenuated less than a triangular one even if both have the same volume when routed through a closed conduit.

A further comparison of the effects the inflow distribution has on the results is shown in Figs. 8.11 to 8.13. The uniform hydrograph takes more time to travel along the conduit (it has a bigger lag time) than the triangular one (both hydrographs attenuated at their peak by 10%). The reason for this is that for a constant inflow factor and a constant peak attenuation the triangular distribution has a much bigger duration than the uniform one. Furthermore, in the case of the triangular distribution, the peak discharge in the outflow hydrograph corresponds to the peak of the inflow hydrograph which lies, in time, in the middle of its duration. In the case of the uniform distribution however, the outflow hydrograph peak will correspond to the inflow peak at a much earlier stage of the distribution, i.e. at the beginning of the inflow hydrographs. This implies a later entry time (in the conduit) for the peak of the uniform distribution resulting in a longer lag time.

The engineer faced with the problems of routing a runoff hydrograph through a pipe or a channel will find the results presented here of direct use. The runoff hydrograph could be the result of overland flow or the outflow from another conduit. Figs. 8.8 to 8.10 can be used to establish the necessity of routing while Figs. 8.11 to 8.13 can be used to calculate a lag time for the cases for which time shift routing is shown to be adequate.

REFERENCES

Argaman, Y., Shamir, U. and Spivak, E. 1973. Design of optimal sewerage systems, Proc. ASCE, (99), EE5, Oct., p 703-716.

Barnes, A.H., 1967. Comparison of computed and observed flood routing in a circular cross section. Intl. Hydrol. Sympos. Colorado State Univ., Fort Collins, pp 121-128.

Constantinides, C.A., 1983. Comparison of kinematic and time shift routing in closed conduits. Report 3/1983. Water Systems Research Programme, University of the Witwatersrand.

Green, I.R.A., 1984. WITWAT stormwater drainage program. Water Systems Research Programme, Report 1/1984. University of the Witwatersrand. 67p

Martin, C. and King, D., 1981. Analysis of storm sewers under surcharge. Proc. Conf. Urban Stormwater, Illinois. pp 74-183.

Merritt, L.B. and Bogan, R.H., 1973. Computer based optimal design of sewer systems. Proc. ASCE, (99), EE1, Feb. pp 35-53.

Stephenson, D., 1980. Direct design algorithm for storm drain networks. Proc. Int. Conf. Urban Storm Drainage, Univ. Kentucky, Lexington.

Stephenson, D., 1981. Stormwater Hydrology and Drainage, Elsevier, 276 pp.

Yen, B.C. and Sevuk, A.S., 1975. Design of storm sewer networks. Proc. ASCE, 101, EE4, Aug. 535-553.

CHAPTER 9

URBAN CATCHMENT MANAGEMENT

EFFECTS OF URBANIZATION

In nature a semi-equilibrium exists between precipitation, runoff and infiltration into the ground. Over years the water table fluctuates about a mean. It recedes during droughts when seepage into watercourse exceeds replenishment rates, and rises when it rains. The depth of soil above the water table is generally not excessive or else vegetation dies, the ground dries out and wind blows the soil away. The amount of water which rises up in the soil under capillary action or in vapour form is limited by the depth of water table.

The construction of impermeable barriers on the surface, such as roads and buildings, reduces the rate of ground water replenishment. The water runs off easier and the limited permeable area restricts infiltration. The groundwater level will therefore drop and the zone above the water table will gradually dry out. Vegetation and the soil characteristics will change. If we are not to affect our environment adversely we should attempt to return some of the stormwater we channel off our urban area back to the ground. This can be accomplished by ensuring adequate permeable surfaces, and by directing stormwater into specially selected or constructed seepage areas. We will then not only maintain the regime but also minimize design flow rates downstream.

The depletion of groundwater will also alter the relationship between rainfall and runoff. After a dry spell more water will be needed to saturate the ground so that the initial abstraction may be greater than before the development occurred. This is offset to an extent by the impermeable ground cover. The net effect is to make a more extreme hydrology i.e. a greater difference between floods and droughts than before development.

Effect on Recurrence Interval

Urban Development affects the rainfall pattern and statistics as well as the runoff pattern. It has been alleged that blanketing effects due to solar shields affect evaporation and hence the resultant precipitation. The blanket of smog, dust, fumes etc., may also affect the place in which the clouds release their moisture, so the effect of urbanization on rainfall is difficult to estimate and the statistical properties of rainfall records (e.g. the mean, coefficient of variance, frequency and distribution) will be affected as well to some extent. Rainfall is reputed to fall more on the leeward side of cities due to the heating up of the air over the city and up to 15% more precipitation has been attributed to this effect. (Huff and Changnon, 1972; Colyer, 1982). Apart from this, the relationship between rainfall and runoff is affected.

Some of the simplistic methods of assessing runoff suppose that the recurrence interval of a calculated flood is the same as the recurrence interval of the causitive rainfall for the design storm duration. It could be that this assumption is borne in mind in the choice of the Rational coefficient. That is the use of the rational method gives a certain recurrence interval of runoff (equal to that of the selected storm in fact) but it does not imply that the design storm is the one which will produce that runoff. This is a gross simplification and it is rarely that the recurrence interval of a storm and its resulting flood coincide. This is due to the predominating effect of abstraction or losses. It will be recalled that generally the Rational coefficient C is nearer 0 than 1 implying losses are greater than runoff. That in turn means that losses, which in turn are mostly soil moisture abstraction, affect runoff more than rainfall. Hence the runoff and its return period should be more related to soil moisture conditions than to rainfall. A study by Sutherland (1982) indicates little correlation between rainfall recurrence interval and the recurrence interval of the flood when assessed in terms of the peak flow rate. He proposed that antecedent moisture conditions, measured in terms of the total precipitation in preceding days, should be a parameter in runoff-duration-frequency relationships. His contention is that the probability of a certain runoff intensity is more related to the probability of the soil being at a certain saturation than the rainfall intensity.

How does urbanization affect the argument? In fact it counters the above ideas. The more the natural surface cover is replaced by impermeable surfaces the more runoff becomes a direct response function to rainfall. In the limit for 100% runoff, soil does not feature and the recurrence interval of runoff is equal to that of the storm causing it.

EXAMPLE:

CALCULATION OF PEAK RUNOFF FOR VARIOUS CONDITIONS

The effect of urbanization on runoff can be illustrated with the following example. In particular it will be seen that the peak flows

174

increase (as well as the volume of runoff).

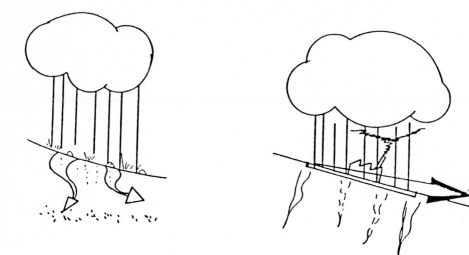

Fig. 9.1.   The effect of urbanization on runoff

i)   **Virgin Catchment**

The simple rectangular catchment depicted in Fig. 9.2 will be studied to indicate the various effects of urban development on the storm runoff peak. The effects computed are reduced roughness, impermeable cover and channelization. A constant frequency, uniform rainfall intensity duration relationship as follows is used:

$$i(mm/h) = \frac{a}{(0.24+t_d)^{0.89}}$$

where t_d is the storm duration in hours.

This is typical of a temperate area, and the value of 'a' for this region is estimated to be 70 mm/h for storms with a 20 year recurrence interval of exceedance.

The catchment is assumed to have a constant slope of 0.01 and initially the cover is grass. The representative Manning roughness for overland flow is estimated to be 0.1. The initial abstraction (surface retention and moisture deficit make up) is 30 mm and subsequent mean infiltration rate over a storm, 10 mm/hr.

Thus $\alpha = \sqrt{S}/n = \sqrt{0.01}/0.1 = 1.0$

Infiltration ratio $F = f/a = 10/70 = 0.143$

Initial loss ratio $U = u/a = 30/70 = 0.429$

Length factor in S I Units $LF = L/36\alpha a^{2/3} = 2000/36 \times 1 \times 70^{2/3} = 3.27$

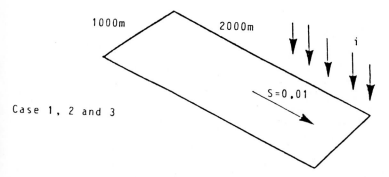

Case 1, 2 and 3

Fig. 9.2   Simple catchment analyzed

From Fig. 3.6 (for U = 0.40) read equilibrium $t_e > 4h$ (off the graph) but the peak runoff factor for this F is QF = 0.23 which corresponds to a storm duration of t_d = 2.2h. The peak runoff rate is

Q_p = $0.23 B \alpha a^{5/3}/10^5$ = $0.23 \times 1000 \times 1 \times 70^{5/3}/10^5$ = $2.74 m^3/s$

The total precipitation rate over the catchment of area A for the same storm duration is

Ai = $\dfrac{70 \times 1000 \times 2000}{(0.24+2.2)^{.89} \times 3600 \times 1000}$ = $17.6 m^3/s$

so the rational coefficient C = 2.74/17.6 = 0.16.

Note however that the full catchment is not contributing at the time of peak runoff for the design storm, so C does not only represent the reduction in runoff due to losses, it also accounts for only part of the catchment contributing. The runoff for the full catchment would be less as the storm duration would be longer than 2.2 h so the intensity would be less and the losses relatively higher.

ii)   **Reduction in Infiltration**

If the infiltration and initial abstractions are reduced by urbanization, the peak runoff increases. The construction of buildings and roads could reduce infiltration rate to 7 mm/h and initial abstraction to 14 mm. For F = 7/70 = 0.1 and U = 14/70 = 0.20 (Fig. 3.5) then for LF = 3.27 as for case (i), the time to equilibrium is off the chart but the critical storm has a duration of 2.2 hours and the corresponding peak flow is

$$Q_p = 0.44 \times 1000 \times 1.0 \times 70^{5\cdot3}/10^5 = 5.24m^3/s$$

The corresponding runoff coefficient C works out to be 0.30

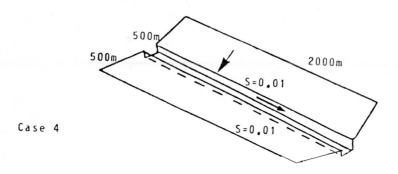

Case 4

Fig. 9.3   Catchment with channel

iii)   Effect of Reduced Roughness due to Paving

With the construction of roads, pavements and building the natural retardation of the surface runoff is eliminated and concentration time reduces. That is, the system response is faster and as a result shorter, sharper showers are the worst from the point of view of runoff peak. For the sample catchment the effective Manning roughness could quite easily be reduced to 0.03. Then $\alpha = 3.33$ and LF = 0.98. The time to equilibrium would therefore be 3h but the peak intensity storm has a duration of 2.2h as before. In this case extent of the storm over the catchment is greater however, and the peak runoff is

$$Q_p = 0.23 \times 1000 \times 3.33 \times 70^{5/3}/10^5 = 9.12m^3/s$$

The corresponding increase in C is from 0.16 to 0.52 an appreciable increase if it is borne in mind this is only due to reduced roughness and does not account for reduced infiltration. It will be noted that the effect of reducing roughness is even greater than decreasing infiltration for this case. The same effect is magnified in the following example.

iv)   Effect of Canalization

The effect of a stream down the centre of the catchment is illustrated in the following example. The same surface roughness (n = 0.1) and permeability (f = 10 mm/h, u = 30 mm) as for case (i) are assumed. The overland flow cross slope is taken as 0.04 and 0.01 for a 8 m wide channel down the catchment. The dimensionless hydrographs in Chapter six are used again.

The stream catchment ratio $G = (\dfrac{2L_s}{b\alpha_s})^{0.6} \dfrac{b\alpha_o^{0.6}}{2L_o}$

$= (\dfrac{2 \times 2000}{8 \times 1})^{0.6} \dfrac{8 \times 2^{0.6}}{2 \times 500} = 0.50$

By trial, guess storm duration resulting in peak runoff of 1.5h, then

$i_e = \dfrac{70}{(0.24 + 1.5)^{.89}} - 10 = 42.7 - 10 = 32.7$ mm/h

$t_{ed} = t_d - t_u = 1.5 - 30/42.7 = 0.80$h

$F = 10/32.7 = 0.31$

$t_{co} = (\dfrac{L_o}{\alpha i_e^{m-1}})^{1/m} = \dfrac{500}{2 \times (32.7/3600000)^{2/3}}^{3/5} \quad 2860s = 0.80$h

$T_D = (5/3)t_{ed}/t_{co} = (5/3)0.8/0.8 = 1.67$

Therefore $t_d = t_{ed} + t_u = 0.8 + 30/42.7 = 1.50$ h which agrees with guess

Interpolating Figs. 6.10 and 6.11 the peak factor Q = 0.85

Peak flow $Q_s = QAi_e = 0.85 \times 2 \times 10^6 \times 32.7/3.6 \times 10^6 = 15.4 m^3/s$

Rational coefficient $C = 15.4/(42.7 \times 2/3.6) = 0.65$

v)   **Combined reduced roughness and reduced losses**

If roughness is reduced by paving to 0.03 then α = 3.33 and LF = 0.98 as for case (iii). The reduced loss factors become F = 0.1 and U = 0.2 as for case (ii). From Fig. 3.5 t_c = 1.7 h and the corresponding QF = 0.43.

Hence the peak flow $Q = 0.43 \times 1000 \times 3.33 \times 70^{53} = 17.0 m^3/s$. The rainfall rate for a storm of this duration is

$\dfrac{70 \times 1000 \times 2000}{(0.24 + 1.7)^{0.89} \times 36000 \times 1000} = 21.6 \ m^3/s$ so $C = 0.79$.

The relative effect of each variable on peak runoff can be compared with the aid of Table 9.1. The effect of reducing infiltration 30% and initial abstraction 40% is to double the peak runoff. The critical storm duration was not affected but the effective area contributing increased slightly. The effect of reducing surface roughness is even more remarkable however. Even maintaining the same losses (both initial and abstraction and infiltration) as for the natural catchment the runoff peak increased by a factor of 4. The area contributing increased noteably although the critical storm duration was not affected. Reducing roughness even more would not necessarily increase runoff much as practically the entire catchment contributes for case (iii) whereas the area contributing in case (i) was much less. Only for case (v) with reduced roughness and losses is the concentration time equal to the critical storm duration.

TABLE 9.1   Showing effect of different surface configurations on peak runoff from a 2000m long by 1000m wide catchment. $S_o = 0.01$, $i = 70$ mm/h/$(0.24h + t_d)^{0.89}$

	CASE	n	f mm/h	u mm	t_c h	t_d h	i mm/h	Q_p m³/s	C
i)	Virgin catchment	0.1	10	30	5	2.2	36.7	2.74	0.16
ii)	Reduced losses	0.1	7	14	4	2.2	36.7	5.24	0.30
iii)	Reduced roughness	0.03	10	30	3	2.2	36.7	9.12	0.52
iv)	Canaliz- ation (stream width 3m)	0.1	10	30	0.8	1.5	42.7	15.4	0.65
v)	Reduced losses and roughness	0.03	7	14	1.7	1.7	38.8	17.0	0.79

The effect of canalization is somewhat similar to reducing roughness - water velocities, and concentration rates, are faster. This is due to the greater depth in channels ($Q = B \sqrt{S} y^{2/3}/n$). Consequently a greater area contributes to the peak.

Not much sense can be made out of comparing the resulting rational coefficients (ratio of peak runoff rate to rainfall rate times catchment area). That is because the time of concentration for each case is different due to differing roughness, rainfall rate etc. In any case it is irrelevant when it comes to critical storm duration which is shorter than the time to equilibrium.

DETENTION STORAGE

Although the kinematic equations as presented previously cannot accommodate reservoir storage they may be rearranged to illustrate the storage components in them. The St. Venant equations which include terms for storage when water surface is not parallel to the bed, are

$$\frac{\partial A}{\partial t} = - \frac{\partial Q}{\partial x} \tag{9.1}$$

$$\frac{\partial v}{\partial t} = g\ (S_o - S_f) - g\ \frac{\partial y}{\partial x} - v\ \frac{\partial v}{\partial x} \tag{9.2}$$

The first equation is the continuity equation and the second the so-called dynamic equation. The first equation does not give the total storage in the reach, it represents the rate of change in cross sectional area of flow as a function of inflow and outflow. The second equation contains more about the distribution of storage. The last two terms represent the wedge component of storage, which are absent in the kinematic equations. The kinematic equations therefore treat storage as a prism, with storage in blocks and no allowance for difference in slope between bed and water surface is made. Since the second equation is replaced by a friction equation and $S_o = S_f$ in the kinematic equations, only the first equation in the case of the kinematic equations can be used to calculate storage changes.

The continuity equation may be written as

$$\frac{O-I}{\Delta x} + \frac{A_2 - A_1}{\Delta t} = 0 \tag{9.3}$$

where O is outflow, I is inflow over a reach of length Δx, and A_1 and A_2 are the cross sectional areas before and after Δt respectively. If $O = (O_1 + O_2)/2$ and $I = (I_1 + I_2)/2$ and $A\Delta x$ is replaced by S, the storage which is a function of A_1 and A_0, which in turn are functions of flowrate, e.g. $S = XI + (1-X)O$, then equation (9.3) becomes the one frequently used for open channel routing,

$$O_2 = c_1 I_1 + c_2 I_2 + c_3 O_1 \tag{9.4}$$

where c_1, c_2 and c_3 are functions of Δx and Δt. The latter equation is referred to as Muskingum's equation used in routing floods along channels. if $X = 0$ the routing equation corresponds to level pool or reservoir routing. The more general equation with $X = 1/2$ represents a 4-point numerical solution of the continuity equation as employed in kinematic models (Brakensiek, 1967).

CHANNEL STORAGE

Channel storage performs a similar function to pond storage in retarding flow, and there are many analogies which can be drawn between the two. Channel storage is a function of friction resistance and channel shape and can be controlled in various ways.

The form of friction equation, as well as the friction factor, affect the reaction speed of a catchment and the volume stored on the catchment. The excess rain stored on the catchment, whether in channels or on planes, is a form of detention storage, and as such, affects the concentration time and consequently the peak rate of runoff. Some friction formulae used in stormwater drainage practice are listed below.

	S.I. units	English units	
Darcy-Weisbach	$Q = (8/f)^{1/2} A(RSg)^{1/2}$	$Q = (8/f)^{1/2} A(RSg)^{1/2}$	(9.5)
Chezy	$Q = 0.55CA(RS)^{1/2}$	$Q = CA(RS)^{1/2}$	(9.6)
Manning	$Q = AR^{2/3}S^{1/2}/n$	$Q = 1.486AR^{2/3}S^{1/2}/n$	(9.7)
Strickler	$Q = 7.7A(R/k)^{1/6}(RSg)^{1/2}$	$Q = 7.7A(R/k)^{1/6}(RSg)^{1/2}$	(9.8)

R is the hydraulic radius A/P where A is the area of flow and P the wetted perimeter. R can be approximated by depth y for wide rectangular channels. S is the energy gradient, f is the friction factor and k is a linear measure of roughness analogous to the Nikuradse roughness.

Both the roughness coefficient α and the exponent m of R or y in the general flow equation (9.11) affect the peak flow off a catchment. This is largely due to the attenuating effect of friction resulting in a larger time to equilibrium. A rainfall excess intensity-duration relationship is required to evaluate the effect of each coefficient on peak runoff rate and maximum catchment storage. The following expression for excess rainfall intensity is assumed:

$$i_e = \frac{a}{(c+t_d)^p} \qquad (9.9)$$

In this equation it is customary to express i_e and a in mm/h or inches per hour and b and t_d in hours where t_d is the storm duration assumed equal to time of concentration t_c for maximum peak runoff of a simple catchment.

Starting with the kinematic equation for continuity

$$\frac{\partial y}{\partial t} + \frac{\partial q}{\partial x} = i_e \qquad (9.10)$$

and a general flow resistance equation

$$q = \alpha y^m \qquad (9.11)$$

then it may be shown that $t_c = (L/\alpha i_e^{m-1})^{1/m}$ where q is the runoff rate per unit width of the catchment and y is the flow depth. The rising limb of the hydrograph is given by the equation

$$q = \alpha (i_e t)^m \qquad (9.12)$$

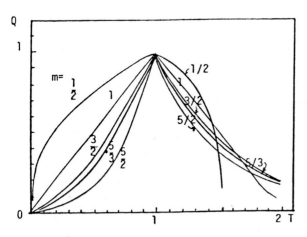

Fig. 9.4   Hydrograph shapes for different values of m in $q = \alpha y^m$

and another expression may be derived from the falling limb (see Chapter 2). In Fig. 9.4 are plotted dimensionless hydrographs to illustrate the effect of m on the shape of the hydrograph. The graphs are rendered dimensionless by plotting $Q = q/i_e L$ against $T = t/t_c$. m is used as a parameter. Thus m = 1/2 represents closed conduit or orifice flow, m = 1 represents a deep vertical sided channel, m = 3/2 represents a wide rectangular channel according to Darcy or a rectangular weir, m = 5/3 represents a wide rectangular channel if Manning's equation is employed, and m = 5/2 represents a triangular weir. The graphs immediately indicate the effect of m on catchment detention storage since the area under the graph represents storage.

The smaller m, the greater storage. Thus provided storage is economical by throttling outflow one may increase storage and increase concentration time thereby reducing discharge rate (which is not immediately apparent from these graphs as they are plotted relative to excess rainfall intensity). In practice the concentration time increases the greater the storage so that the lower intensity storms become the design storms. This has a compound effect in reducing flow rates since total volume of losses increases and it is possible that the entire catchment will not contribute at the peak flow time.

A general solution of peak flow and storage in terms of intensity-duration relationships is derived below. Solving (9.9) with $t_d = t_c$ for maximum rate of runoff per unit area and generalizing by dividing by a,

$$q_m/aL = i_e/a = \cfrac{1}{\{c + \cfrac{(L/\alpha i_e^{m-1})^{1/m}}{3600}\}^p}$$

$$\cfrac{1}{\{c + \cfrac{\left|L/\alpha(a/3600000)^{m-1}\right|^{1/m}}{3600(i_e/a)^{1-1/m}}\}^p} \qquad (9.13)$$

The term $L/\alpha a^{m-1}$ is referred to as the length factor. The constants are introduced for a in mm/h, and time of concentration in hour units. The maximum peak flow factor i_e/a is plotted against length factor in Fig. 9.5, since it is not easy to solve (9.13) directly for i_e/a

i_e/a and s/a

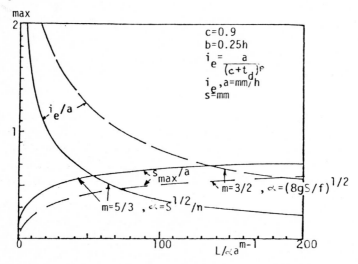

Fig. 9.5 Peak flow and storage versus length factor

An expression for the corresponding catchment storage is derived below. At equilibrium the flow per unit width at a distance x down the catchment is

$$q = i_e x$$
$$= \alpha y^m$$

therefore $\quad y = (i_e x / \alpha)^{1/m}$

Integrating y with respect to x yields the total volume on the catchment

$$V = \frac{Lm}{m+1} (i_e L / \alpha)^{1/m}$$

or in terms of the average depth of storage $\quad s = V/L$

$$s/a = \frac{m}{m+1} (\frac{i_e}{a})^{1/m} (\frac{L}{\alpha (a/3600000)^{m-1}})^{1/m} \frac{1}{3600} \qquad (9.14)$$

where s is in mm, and i_e and a are in mm/h. s/a is also plotted against length factor in Fig. 9.5. It will be observed that average storage depth does not increase in proportion to $L/\alpha a^{m-1}$. In fact the rate of increase reduces beyond $L/\alpha a^{m-1} = 50$, and the rate of reduction in peak flow i_e/a also decreases beyond the figure, indicating reducing advantage in increasing channel length or roughness ($\alpha = K_1 \sqrt{(S)} /n$). Since total channel cost is a direct function of storage capacity it would appear to be an optimum at some intermediate value of $L/\alpha a^{m-1}$ if there is a cost associated with peak discharge e.g. culverts or flooding downstream (see Fig. 9.6).

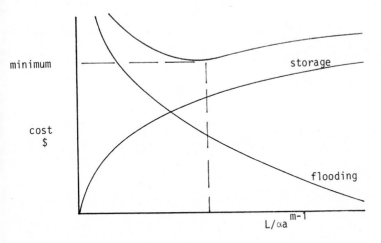

Fig. 9.6   Optimum catchment storage volume.

Note that infiltration after the rainfall stops, is neglected in the above analysis. Inclusion of that effect would lower the i_e/a and s/a lines to the right, implying a larger $L/\alpha a^{m-1}$ is best. The model provides an indication of total storage in the system. The location (and volume) of storage could be further optimized using dynamic programming methods or by detailed modelling. It should be found generally that it is most economical to provide pond storage ($m = 1/2$) at the outlet, whereas channel or catchment storage ($m = 5/3$) is most economical at the head of the system.

KINEMATIC EQUATIONS FOR CLOSED CONDUIT SYSTEMS

If the open channel kinematic equations are applied to closed conduit flow the problem becomes a steady state flow one since flow rates become independent of cross section. This is provided the conduits remain full and there are no storage ponds at nodes joining conduits. If one permits storage variation at nodes one has the reservoir-pipe situation encountered in water supply which is often analyzed employing pseudo-steady flow equations.

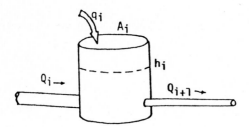

Fig. 9.7   Input–output node storage

The continuity equation becomes (see Fig. 9.7)

$$(Q_{i+1} - Q_i) - q_i + A_i \frac{dh_i}{dt} = 0 \qquad (9.15)$$

where the reservoir surface area A_i replaces $B\ dx$ in the open channel continuity equation where B is the catchment width. q is the reservoir inflow here. The dynamic equation is replaced by

$$Q_i = \alpha A^m \qquad (9.16a)$$

where A is the (constant) conduit cross sectional area. Since the kinematic equations omit the dependency of Q on head difference h, the latter equation assumes the head gradient along the pipe equals the pipe gradient, i.e. free-surface just full flow. Since A is a constant it is relatively easy to replace the last equation by one of the form

$$Q_i = \alpha A h_i^m \qquad (9.16b)$$

This equation is applicable to free discharge from an orifice or over a weir. One more applicable to conduit flow would be

$$Q = \alpha A (h_{i-1} - h_i)^m \qquad (9.16c)$$

Any one of the above three equations could be applicable in storm-water drainage. For channel or overland flow (9.16a) applies, for complete storage control (9.16b) applies and for closed conduit control (9.16c) is applicable. The latter form of equation has in fact been employed in water reticulation pipe network analysis. It can be applied in storm drainage to closed systems (not of great interest in stormwater management practice) or to pipe-reservoir problems. Surface detention and artificial detention storage ponds can be handled in an overall flow balance employing the closed conduit kinematic method. It should be noted that the numerical instability problems associated with solution of the open channel kinematic equations are absent. Time steps can be much larger than for open channel kinematic modelling. Storage fluctuations may be computed in steps and the effect of changes in pond water levels on flows in conduits can be accounted for.

One possible application of such a program is to an inter-connected pond system with reversible flows in conduits. Overload from one pond can be forced back to another pond. Such situations can readily arise from spatially variable storms and possibly for travelling storms.

Off-channel storage can also be accounted for. Such ponds have the advantage that water level variations are not as marked as the head variations in the drain pipes (which may in fact be surcharged). This is due to the reversible head loss between the main conduit and the pond.

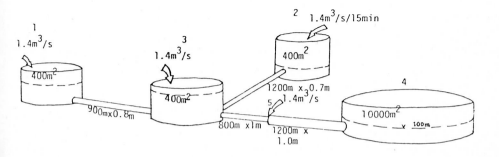

Fig. 9.8   Conduit and storage storm drain network.

The simplified layout in Fig. 9.8 was analyzed employing the accompanying kinematic closed conduit continuous simulation program. Input and output are appended to illustrate the simplicity in this type of anlysis. Flow reversal, pond level variations and the large attenuation in peak flow will be observed due to the ponds (from $5.6m^3/s$ down to $1.5m^3/s$). By adjusting individual pond areas and conduit sizes an optimum design could be achieved for any design storm input. A sensitivity analysis for alternative storms such as different storm durations or ones with spatial variability would then be performed.

COMPUTER PROGRAM TO SIMULATE RESERVOIR LEVEL VARIATIONS IN A PIPE NETWORK

Closed conduit drainage networks can as explained above, be used to ameliorate peak flows by directing water into storage. Flow can be in either direction and depends on the difference in water levels at the two ends of the conduit, not on the conduit gradient as for open channels. Apart from this, the principles are the same as for open channel kinematic flow. That is steady state conditions (head loss/flow equations) are used together with the continuity equations. The accompanying computer program written in HP 85 'BASIC' will simulate the variations in water level in reservoirs in addition to performing a network flow balance.

The program is based on the linear node method (Stephenson, 1984) network analysis with an additional variable, area of reservoir for each 'fixed head' or, in this case, 'reservoir type' node. If the simulation duration T4 in hours and time increment T5 are input, for example 24 and 1, then the heads at each node and water level in each reservoir will be printed out every hour. The actual network iterations each time interval after the first should be minimal since the network flows are balanced in the first iteration and only unbalanced due to reservoir level changes which will have to be corrected at subsequent time intervals. Although drawoffs are time-fixed in the present program, they could be altered at pauses in the running or inserted in equation form.

The output, namely level variations, could be used to estimate required reservoir depths (using trial reservoir surface areas) and in fact to see at which reservoir locations the storage is most required. Data requirements are similar to the analysis program with the following additions.

In the first data line after the name, the simulation duration and increment in hours is added at the end of the line. In the pipe data, the first pipes should be from the various reservoirs with the surface areas of the up-stream reservoirs in square metres given at the end of the pipe data lines. In order to display the reservoir levels in the biggest reservoir it is necessary to have a supply pipe from a pseudo fixed head, very large, reservoir to represent a pumped supply feeding into the actual biggest level reservoir in the distribution system.

The selection of 'upper' and 'lower' nodes for any pipe, numbered 1 and 2 is somewhat arbitrary. If the incorrect flow direction is assumed, a negative flow number will appear in the answers thus indicating the flow direction is from node 2 to node 1 as specified.

When data is put in, the order of pipes is to a limited extent arbitrary, but the 'node 1' of any pipe should have been defined as a 'node 2' in some previous pipe. This does not apply to the first pipe which will originate at a reservoir. The order of pipes enables data on successive nodes, i.e. initial estimates of heads and flows, to proceed down the system from previously defined nodes.

Node numbering is also open to the user except the reservoir-type nodes (with specified initial water levels) should be numbered first, from 1 to J3.

There is scope for setting all Darcy friction factors the same to minimize data requirements, or to vary each factor. Note if old data in files is used those friction factors, not the 'common' factor, is used even if a common factor is fed in. To print out 'old data' in file, it is necessary to go into revision mode (2) of pipe data input. To get out of revision mode, type 0 for pipe number to be revised.

Part of the data is read interactively on the keyboard. The first lines of data (name, duration, no. of nodes, reservoir data and pump data) is typed for each run. The pipe and node data can be typed in or retrieved from a file or ammended in a file.

The time increment between iterations for simulation mode must be small enough to avoid large variations in water levels in reservoirs between iterations. The reservoir surface area and flows will control this.

Additional pipes can be added in edit (2) mode and will then be stored in the data file. Pipes can only be removed by limiting the number of pipes in the initial lines to eliminate those not required at the end. The other way is to put a very small diameter for a pipe to be removed from the network. New nodes or reservoirs can be added by retyping in data.

When reading in initial data however, no more than the number of pipes in the data file should be specified. The number will automatically be increased when more data lines are added.

The last specification of any drawoff is retained if a node happens to be specified more than once in input. One should also make sure each node is specified (as a N2) at least once to define its drawoff.

Data Input

Each line may contain more than one unit of data separated by commas.

Line 1	Name of network (and run no.)
Line 2	Analysis (0) or simulation (1) – type 0 or 1
Line 3	Drawoff duration in minutes, thus if drawoff is over 8 hours, type 480. Simulation duration mins. If 24 hours, type 1440, Time increment DT, mins. Suggest 30 – 120.
Line 4	Constant (0) or various (1) Darcy f's – type 0 or 1
Line 5	No. pipes,
	No. nodes (total including reservoirs)
	No. reservoir type nodes.
Lines 6...	(one for each reservoir node in successive order)
	Initial water level, m
	Surface area of reservoir, m²
Line 7	Old (0) or new (1) or revised pipe data (2);
	type 0,1 or 2.
Lines 8...	(one for each pipe in new pipe data)
	Node 1 no.
	Node 2 no.
	Pipe length m
	Pipe inside dia., m
	Drawoff at node 2, m³/s
	(Darcy friction factor if line 4 is 1)
Line 9	If line 7 is 2, will ask pipe no. for revision.
Line 10	Pipe data for new pipes as for Line 8 including Darcy friction factor.
Line 11	No. of pumps or pressure reducing valves (one per pipe).
Lines 12...	Pipe no. in which pump or PRV is installed, pumping head or PRV head loss (–) in m.

List of Symbols in Program

A1	1 = analysis, 2 = simulation
A2	0 = constant f, 1 = varying Darcy f.
A4	0 = old data, 1 = new data, 2 = revise old data
A5	0 = no data listing required, 1 = required
C(K)	headloss/Q|Q|
C2	ΣH for each SOR
C3	$\Sigma \Delta$F
D(K)	pipe diameter (m)
D2	old value of H(I)
F(K)	Darcy friction factor e.g. 0.012 large dia. clean pipe
	0.03 small tuberculated pipe
F1	common Darcy factor
H(I)	head at node or junction I
I	node counter
J	number of nodes
J1(I)	upper node number of pipe
J2(I)	lower node number on pipe
J3	number of reservoir type nodes
K	iteration
K1	pipe counter
L	node counter
M	pipe counter
M1(L)	number of connecting pipes
M2(L,M1(L))	pipe number connecting
N$	alphanumeric name of system, up to 12 characters
NO	maximum number main iterations permitted e.g. $\sqrt{J}$ + 5
N1	maximum number SOR (successive over-relaxation of simultaneous equations) iterations e.g. $\sqrt{J}$ + 10
N2	counter for main iterations
N3	counter for SOR iterations
P	number of pipes
P1	numer of pipes and PRV's (1 per pipe maximum)
Q(K)	flow in pipe
Q1	drawoff m^3/s
Q2(I)	drawoff m^3/s
R(k)	pump head in m, (or pressure reducing valve head in m if negative)
S	gπ^2/8

S(2)I	ΣKij
S3	ΣHj
S4(I)	$\Sigma KijHj$
S5	old Q(K) for averaging
T3	drawoff duration, mins e.g. 8h × 60 = 480
T4	simulation duration, mins e.g. 24 × 60 = 1440
T5	time increment in simulation, mins e.g. 60
T0	tolerance on head in m e.g. 0.0001
T1	tolerance on SOR in m e.g. 0.01
W-SOR	factor e.g. 1.3 (1-2)
X(K)	pipe length m

REFERENCES

Brakensiek, D.L., 1967. Kinematic flood routing. Trans Am. Soc. Agric. Engrs. 10(3) p 340-343.

Colyer, P.J., 1982. The variation of rainfall over an urban catchment. Proc. 2nd Intl. Cong. Urban Storm Drainage. University of Illinois.

Huff, F.A. and Changnon, S.A., 1972. Climatological assessment of urban effects on precipitation at St. Louis. J. Appl. Meteorology, 11, p 823-842.

Stephenson, D., 1984. Kinematic analysis of detention storage. Proc. Storm Water Management and Quality users Group Meeting, USEPA, Detroit.

Stephenson, D., 1984. Pipeflow Analysis, Elsevier, Amsterdam, 204 pp.

Sutherland, F.R., 1983. An improved rainfall intensity distribution for hydrograph synthesis. Water Systems Research Programme, Report 1/1983, University of the Witwatersrand.

Program Listing

```
10 ! NETSIM KINEMATIC CONTIN SI
   MULN OF NETWORKS WITH STORAG
   E
20 ASSIGN# 1 TO "DATNET" ! CREA
   TE"DATNET",100,88
30 DIM C(90),Q(90),H(90),Q2(90)
   ,S2(90),S4(90),F(90)
40 DIM J1(90),J2(90),D(90),X(90
   ),R(90),M1(90),M2(90,9)
50 DISP "NAME OF NETWORK";
60 INPUT N$
70 DISP "ANALYSIS OR SIMULATION
   (1/2)";
80 INPUT A1
90 IF A1=2 THEN 140
100 T3=1
110 T4=1
120 T5=1
130 GOTO 160
140 DISP "DRAWOFF DURATIONmin,SI
    M DURNmin,DTmin";
150 INPUT T3,T4,T5
160 Q2(1)=0
170 T3=T3*60
180 T4=T4*60
190 T5=T5*60
200 DISP "VARYING fs(0/1)";
210 INPUT A2
220 IF A2=0 THEN 260
230 DISP "NPIPES,NODES,NRESS";
240 INPUT P,J,J3
250 GOTO 280
260 DISP "NPIPES,NODES,NRESS,DAR
    CYf";
270 INPUT P,J,J3,F1
280 DISP "INITL WATER LEVELm,SUR
    FACE AREA m2";
290 FOR L=1 TO J3
300 DISP L;
310 INPUT H(L),A(L)
320 NEXT L
330 G=9.8
340 S=3.14159^2*G/8
350 DISP "OLD OR NEW OR REVISE P
    IPEDATA(0/1/2)";
360 INPUT A4
370 IF A4=1 THEN 430 ! NEW DATA
380 FOR K=1 TO P ! OLD DATA
390 READ# 1,K ; J1(K),J2(K),X(K)
    ,D(K),Q2(J2(K)),F(K)
400 NEXT K
410 IF A4=2 THEN 580
420 GOTO 740
430 IF A2=0 THEN 460
440 DISP "NODE1,NODE2,Lm,Dm,DRAW
    OFF2m3/s,DARCYf";
450 GOTO 470
460 DISP "NODE1,NODE2,Lm,Dm,DRAW
    OF2m3/s";
470 FOR K=1 TO P ! PIPE DATA

480 DISP K;
490 IF A2=0 THEN 520
500 INPUT J1(K),J2(K),X(K),D(K),
    Q1,F(K)
510 GOTO 540
520 INPUT J1(K),J2(K),X(K),D(K),
    Q1
```

```
540 Q2(J2(K))=Q1
550 PRINT# 1,K ; J1(K),J2(K),X(K
    ),D(K),Q2(J2(K)),F(K)
560 NEXT K
570 GOTO 740
580 FOR K1=1 TO 100
590 DISP "REVISE PIPE NO.";
600 INPUT K
610 IF K=0 THEN 660
614 IF K<=P THEN 620
616 P=K
620 DISP "NODE1,NODE2,Lm,Dm,DRAW
    OFF2m3/s,DARCYf";
630 INPUT J1(K),J2(K),X(K),D(K),
    Q2(J2(K)),F(K)
640 PRINT# 1,K ; J1(K),J2(K),X(K
    ),D(K),Q2(J2(K)),F(K)
650 NEXT K1
660 DISP "DATALIST REQD (0/1)";
670 INPUT A5
680 IF A5=0 THEN 740
690 PRINT "NODE1 N2 Xm   Dm  Qm3
    /s f"
700 FOR K=1 TO P
710 PRINT USING 730 ; J1(K),J2(K
    ),X(K),D(K),Q2(J2(K)),F(K)
720 NEXT K
730 IMAGE DDDD,DDDD,DDDDD,D.DDD,
    D.DDD,D.DDD
740 FOR K=1 TO P
750 Q(K)=3.14159*D(K)^2/4
760 R(K)=0
770 C(K)=S*D(K)^5/F(K)/X(K) ! 1/
    K
780 IF J2(K)<=J3 THEN 800
790 H(J2(K))=H(J1(K))-1/C(K)*Q(K
    )^2
800 NEXT K
810 DISP "NO.PUMPS/PRVs";
820 INPUT P1
830 FOR P2=1 TO P1
840 DISP "PIPEN,+HEADm N1-N2";P2
    ;
850 INPUT K,R(K)
860 NEXT P2
870 FOR L=1 TO J
880 M1(L)=0
890 FOR M=1 TO P
900 IF J1(M)=L THEN 920
910 IF J2(M)<>L THEN 940
920 M1(L)=M1(L)+1
930 M2(L,M1(L))=M
940 NEXT M
950 NEXT L
960 W=1.3 ! SOR FACTOR
970 T0=.0001 ! TOLERANCE m
980 T1=.01 ! SOR TOL m
990 N0=SQR(J)+5 ! ITNS PIPES
1000 N1=SQR(J)+10 ! ITNS SOR
1010 N2=0
1020 N3=0
1030 PRINT "PIPENET",N$
1040 FOR T6=T5 TO T4 STEP T5
1050 IF T6<=T3 THEN 1090
1060 FOR L=1 TO J
1070 Q2(L)=0
1080 NEXT L
1090 FOR I=1 TO N0
```

```
1110 FOR K=1 TO J ! NEW H BY SOR
1120 S2(K)=0
1130 NEXT K
1140 FOR K=1 TO P
1150 S2(J1(K))=S2(J1(K))-C(K)/AB
     S(Q(K))
1160 S2(J2(K))=S2(J2(K))-C(K)/AB
     S(Q(K))
1170 NEXT K
1180 FOR K=1 TO N1
1190 C2=0
1200 S3=0
1210 N3=N3+1
1220 IF J3+1>J THEN 1380
1230 FOR L=J3+1 TO J
1240 S4(L)=0
1250 FOR M3=1 TO M1(L)
1260 M=M2(L,M3)
1270 IF J1(M)<>L THEN 1290
1280 S4(J1(M))=S4(J1(M))+C(M)/AB
     S(Q(M))*(H(J2(M))-R(M))
1290 IF J2(M)<>L THEN 1310
1300 S4(J2(M))=S4(J2(M))+C(M)/AB
     S(Q(M))*(H(J1(M))+R(M))
1310 NEXT M3
1320 D2=H(L)
1330 H(L)=H(L)*(1-W)+W*(Q2(L)-S4
     (L))/S2(L)
1340 C2=C2+ABS(H(L)-D2)
1350 S3=S3+1
1360 NEXT L
1370 IF C2/S3<=T1 THEN 1390
1380 NEXT K
1390 FOR K=1 TO P ! NEW FLOWS
1400 S4=Q(K)
1410 Q(K)=C(K)/ABS(Q(K))*(H(J1(K
     ))-R(K)-H(J2(K))+.01)
1420 IF I=1 THEN 1440
1430 Q(K)=.5*(Q(K)+S4)
1440 NEXT K
1450 C3=0 ! TOLERANCE CHECK
1460 FOR K=1 TO P
1470 C3=C3+ABS(F(K)-ABS(H(J1(K))
     +R(K)-H(J2(K)))/Q(K)^2*C(K)
     *F(K))

1480 NEXT K
1490 IF C3/P<=T0 THEN 1510
1500 NEXT I
1510 FOR L=1 TO J3 ! RES LEVELS
1520 H(L)=H(L)-Q2(L)*T5/A(L)
1530 FOR M3=1 TO M1(L)
1540 M=M2(L,M3)
1550 IF J1(M)<>L THEN 1570
1560 H(L)=H(L)-Q(M)*T5/A(L)
1570 IF J2(M)<>L THEN 1590
1580 H(L)=H(L)+Q(M)*T5/A(L)
1590 NEXT M3
1600 NEXT L
1610 PRINT USING "K,DDDDDD,X,K,D
     DDDD.DD" ; "Ts=",T6,"H1=",H
     (1)
1620 PRINT "NODE1 N2   Xm    Dm   Q
     m3/s    H2m "
1630 FOR K=1 TO P
1640 PRINT USING 1660 ; J1(K),J2
     (K),X(K),D(K),Q(K),H(J2(K))
1650 NEXT K
1660 IMAGE DDD.DDD,DDDDD,DD.DDD,
     DDD.DDD,DDDDD.D
1670 NEXT T6
1680 STOP
1690 ASSIGN# 1 TO *
1700 END
```

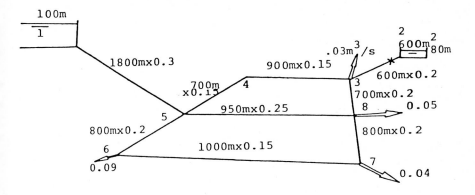

Fig. 9.9   Pipe network analyzed

<table>
<tr><td>NAME OF NETWORK?</td></tr>
</table>

```
NAME OF NETWORK?
TESTSIM 1
ANALYSIS OR SIMULATION(1/2)?
2
DRAWOFF DURATIONmin,SIM DURNmin,
DTmin?
480.960 240
VARYING fs(0/1)?
0
NPIPES,NODES,NRESS,DARCYf?
9,8,2,.02
INITL WATER LEVELm,SURFACE AREA
m2 1 ?
100,1000000
  2 ?
80,400

OLD OR NEW OR REVISE PIPEDATA(0/
1/2)?
1
NODE1,NODE2,Lm,Dm,DRAWOF2m3/s 1
?
1,5,1800,.3,0
  2 ?
5,6,800,.2,.09
  3 ?
5,8,950,.25,.05
  4 ?
6,7,1000,.15,.04
  5 ?
8,7,800,.2,.04
  6 ?
5,4,700,.15,0
  7 ?
4,3,900,.15,.03
  8 ?
8,3,700,.2,.03
  9 ?
3,2,600,.2,0
NO.PUMPS/PRVs?
1
PIPEN,+HEADm N1-N2 1 ?
1,1
```

```
PIPENET              TESTSIM 1
Ts=14400 H1=  100.00
NODE1 N2   Xm    Dm   Qm3/s   H2m
   1   5 1800   .300   .143   74.0
   5   6  800   .200   .076   50.1
   5   8  950   .250   .069   66.3
   6   7 1000   .150  -.014   54.3
   8   7  800   .200   .054   54.3
   5   4  700   .150   .010   72.5
   4   3  900   .150   .010   70.7
   8   3  700   .200  -.035   70.7
   3   2  600   .200  -.055   78.0
Ts=28800 H1=  100.00
NODE1 N2   Xm    Dm   Qm3/s   H2m
   1   5 1800   .300   .146   73.1
   5   6  800   .200   .076   49.0
   5   8  950   .250   .070   65.1
   6   7 1000   .150  -.014   53.2
   8   7  800   .200   .054   53.2
   5   4  700   .150   .010   71.4
   4   3  900   .150   .010   69.2
   8   3  700   .200  -.034   69.2
   3   2  600   .200  -.053   76.1
Ts=43200 H1=  100.00
NODE1 N2   Xm    Dm   Qm3/s   H2m
   1   5 1800   .300  -.010   98.2
   5   6  800   .200   .009   97.9
   5   8  950   .250   .037   95.9
   6   7 1000   .150   .009   96.2
   8   7  800   .200  -.008   96.2
   5   4  700   .150   .017   93.9
   4   3  900   .150   .017   88.3
   8   3  700   .200   .046   88.3
   3   2  600   .200   .063   78.4
Ts=57600 H1=  100.00
NODE1 N2   Xm    Dm   Qm3/s   H2m
   1   5 1800   .300  -.017   99.4
   5   6  800   .200   .009   99.0
   5   8  950   .250   .036   97.2
   6   7 1000   .150   .008   97.5
   8   7  800   .200  -.008   97.5
   5   4  700   .150   .016   95.3
   4   3  900   .150   .016   90.0
   8   3  700   .200   .045   90.0
   3   2  600   .200   .061   80.6
```

CHAPTER 10

KINEMATIC MODELLING

INTRODUCTION

Kinematic flow holds for those cases when a unique relationship exists between the depth of flow and the volumetric flowrate. Model equations are derived through simplifications to the full equations governing gradually-varied, unsteady overland and open channel flow. When applied to one-dimensional overland flow, if the rainfall rate is steady and the watershed geometry is a regular geometry such as a plane, analytical solutions can be obtained for the equivalent character-istic form of the governing equations. Otherwise, one must use numerical solution techniques. Currently, kinematic models best apply to highly impervious (urban) and/or small watersheds. However, research is on-going in several countries to extend the applicability to large and multiple land use watersheds.

The study of kinematic hydrology and modelling must begin with the derivation of the full equations governing overland and open channel flow, followed with an examination of model simplifications and when they can be invoked, development of the characteristic roots, and then proceed with analytical and numerical solutions and example applications. In this chapter, a discussion is given of general modelling concepts and definitions to provide insights and understanding of the role of kinematic modelling as one approach to hydrologic modelling.

STORMWATER MODELLING

Kinematic modelling falls under the umbrella of stormwater modelling. Stormwater is defined as the direct watershed response to rainfall (Overton and Meadows, 1976). It is the runoff which enters a ditch, stream or storm sewer which does not have a significant base flow component. This definition does not assume that all stormwater reaches an open channel by the overland flow route, although in urban areas the direct response is mostly through overland flow due to the high degree of imperviousness. In contrast, in rural watersheds, an overland flow component may be nonexistant and direct storm response may be only near the stream and occur as shallow subsurface flow.

As defined, stormwater is associated with small upland or headwater watersheds where base flow is not a significant portion of the total streamflow during periods of rainfall. Therefore, the emphasis of stormwater modelling is on the storm hydrograph and not the streamflow hydrograph.

MATHEMATICAL MODELS

A mathematical model is simply a quantitative expression of a process or phenomenon one is observing, analyzing, or predicting. Since no process can be completely observed, any mathematical expression of a process will involve some element of stochasticism, i.e. uncertainty. Hence, any mathematical model formulated to represent a process or phenomenon will be conceptual to some extent and the reliability of the model will be based upon the extent to which it can be or has been verified. Model verification is a function of the data available to test the model scientifically and the resources available (time, manpower, and money) to perform the tests. Since time, manpower, and money always have finite limits, decisions must be made as to the degree of complexity the model is to have, and the extensiveness of the verification tests that are to be performed.

The initial task of the modeler then is to make decisions as to which model to use or to build, how to verify it, and how to determine its statistical reliability in application, e.g., feasibility, planning, design, or management. This decision-making process is initiated by clearly formulating the objective of the modelling endeavour and placing it in the context of available resources.

If the initial model form does not achieve the intended objective, then it simply becomes a matter of revising the model and repeating the experimental verifications until the project objective is met. Hence, mathematical modelling is by its nature heuristic and iterative. The choice of model revisions as well as the initial model structure will also be heavily affected by the range of choice of modelling concepts available to the modeler, and by the skill which the modeler has or can develop in applying them.

Figure 10.1 is a schematic representation of the modelling process. The modelling process is not new but is nothing more than a modern expression of the classical scientific thought processes involved in the design of an experiment. What is new is that today a very large number of concepts can be evaluated efficiently in a very small amount of time at a relatively small expense using computers and the body of analytical techniques termed systems analysis.

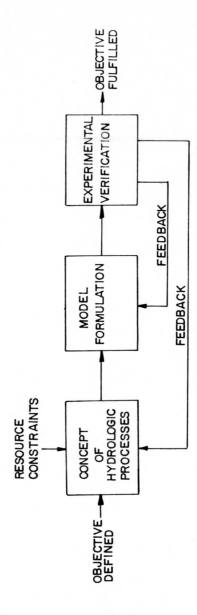

Fig. 10.1 The modelling process

SYSTEM DEFINITION

Dooge (1976) has developed a good working definition of a system as being any structure, device, scheme, or procedure that interrelates an input to an output in a given time reference.

The key concepts of a system are:

1. A system consists of parts connected together in accordance with some sort of plan, i.e. it is an ordered arrangement.
2. A system has a time frame
3. A system has a cause-effect relation.
4. A system has the main function to interrelate an input and output, e.g., storm rainfall and storm runoff.

In the strictest definition, the systems approach is an overall one and does not concern itself with details which may or may not be important and which, in any case, may not be known. This seemingly limits the systems approach to an attempt to get around the complex geometry and physics of the hydrologic system. If we were solely concerned with problems of identification (defined by Doodge as the recognition of the overall nature of a system's operation, but not any details of the nature of the system itself), this attitude of ignoring the details of the system would be a reasonable one. However, when we are going to simulate a hydrologic system and its response, the elements of *physical* hydrology become important. For instance, if we build or use a model that is in conflict with the physical realities, then we can hardly expect to obtain good results from such a model, or even to be able to calibrate the model to achieve good results. Thus, the systems approach to stormwater modeling must consider the assimilation of process models into an overall representation of the hydrologic cycle, or portions thereof. How well the model components must represent the different processes depends on the purpose for the model and how much data are available with which to verify the model.

In conclusion, the essence of systems analysis as applied to stormwater modelling is to interrelate rainfall (input) to stormwater (output) with a reliable model in a computationally efficient manner.

TERMINOLOGY AND DEFINITIONS

There has been an evolution of systems jargon, and it is important to review the main parts to better understand hydrologic modelling.

A *variable* has no fixed value (e.g., discharge) whereas a *parameter* is a constant whose value varies with the circumstances of its application (e.g., Manning n-value).

The distinction between *linear* and *nonlinear* systems is of paramount importance in understanding the mechanism of hydrological modelling. A linear system is defined mathematically by a linear differential equation, the principle of superposition applies and system response is only a function of the system itself. An example of a linear system representation is the unit hydrograph model. A nonlinear system is represented by a nonlinear differential equation and system response depends upon the system itself and the input intensity. An example of a nonlinear system representation is kinematic overland flow. It is well known that real world systems are highly nonlinear, but linear representations have often been made because the system is not under-stood well or because of the pressures exerted by resource constraints.

The *state* of a system is defined as the values of the variables of the system 'at an instant in time. Hence, if we know exactly where all of the stormwater is and its flowrate in a basin, then we know the state of the system. The state of a stormwater system is determined either from historical data or by assumption.

System *memory* is the length of time in the past over which the input affects the present state. if stormwater from a basin today is affected by the stormwater flow yesterday, the system (watershed) is said to have a finite memory. If it is not affected at all, the system has no memory; and, if it is affected by storm flows since the beginning of the world, the system is said to have infinite memory. Memory of surface water flow systems is mostly a function of antecedent moisture conditions.

A *time-invariant* system is one in which the input-output relation is not dependent upon the time at which the input is applied to the system. To illustrate, unit hydrograph models represent the catchment as a time-invariant system because the same unit hydrograph (response function) is maintained throughout the storm regardless of variations in watershed conditions. Usually, time-variance is considered among storm events, seasons of the year, etc., and not within individual storms. Time-invariance indicates constant land use, ground cover, and drainage system configuration and capacity, and ignores soil moisture variations and the effects of erosion.

A *lumped* variable or parameter system is one in which the variations in space either do not exist or have been ignored. Conversely, a *distributed* parameter system recognizes spatial variations. The input is said to be lumped if rainfall into a system is considered to be spatially uniform. Lumped systems are represented by ordinary differential equations and distributed systems are represented by partial differential equations.

A system is said to be *stochastic* if for a given input there is an element of chance or probability associated with obtaining a certain output. A *deterministic* system has no element of chance in it, hence for a given input a completely predictable output results for given initial and/or boundary values. A *purely random process* has no deterministic component and output is completely given to chance. A parametric or conceptual model does have an element of chance built into it since there alway will be errors in verifying it on real data. It does therefore have a stochastic component. A *black box* model relates input to output by an arbitrary function, and has no inherent physical significance.

Model *optimization* is the objective determination of the "best" values for the model parameters using hydrologic data for the type of watersheds and range of hydrologic conditions for which the model has been designed. This function is limited to parametric stormwater models, and is applied in the regionalization process. To *regionalize* a model means to develop a scientific basis for predicting the model parameters on ungauged watersheds from hydrologic and physiographic characteristics of that watershed. Regionalization can be accomplished only if there are enough benchmark watersheds with adequate periods of record that a statistical inference can be drawn, i.e., statistically significant parameter prediction equations can be developed.

Model *calibration* basically is the fine-tuning of model parameter values to achieve the best fit between observed and predicted runoff hydrographs. To *verify* a model is to compare model predictions with observed runoff values without adjusting parameter values to confirm the model is doing a reasonable job in simulating the true watershed response to known input.

Two concepts that are frequently confused (misused) are *analysis* and *simulation*. The confusion with analysis stems from what it is being used to describe. As it relates to stormwater models, analysis is the procedure used to calibrate a model to the data. It is an attempt to improve the state-of-the-art and is fundamentally a research and development tool. Simulation, by contrast, utilizes the results of previous analyses (and regionalization methods) to synthesize (predict) stormwater runoff from either design or real time rainfall on ungauged watersheds. Simulations also can be performed at gauged watersheds to generate runoff data for design events or events not contained in the available record. Analysis is often applied, for example, in the context of studying the probable performance of a storm sewer system during design storm events. We are prone to say that we have analyzed the system and found that it should work! Actually, what we are doing is using simulation results to predict the probable performance characteristics of the storm sewer system.

MODELLING APPROACHES

There are two conceptual approaches that have been used in developing stormwater models. An approach often employed in urban planning has been termed *deterministic* modeling or system simulation. These models have a theoretical structure based upon physical laws and measures of initial and boundary conditions. When conditions are adequately specified, the output from such a model should be known with a high degree of certainty. In reality, however, because of the complexity of the stormwater flow process, the number of physical measures required would make a complete model intractable. Simplifications and approximations must therefore be made. Since there are always a number of unknown model coefficients and parameters that cannot be directly or easily measured, it is required that the model be verified. This means that the results from usable deterministic models must be verified by being checked against real watershed data wherever such a model is to be applied.

The second conceptual stormwater modeling approach has been termed *parametric* modelling. In this case, the models are somewhat less rigorously developed and generally simpler in approach. Model parameters are not necessarily defined as measurable physical entities although they are generally rational. Parameters for these models are determined by fitting the model to hydrologic data with an optimization technique. Application of parametric models to ungauged watersheds is possible only if regionalized parameter prediction equations are available and are based on data from watersheds within the same geographical area and with similar geomorphic and land use characteristics as the watershed being considered. As with deterministic models, user confidence stems from verification studies using local data.

EXAMPLES OF PARAMETRIC AND DETERMINISTIC MODELS

An excellent example of a parametric stormwater model is the TVA Stormwater Model (Betson, et al., 1980). The model is an event simulation model formulated with a variation of the SCS curve number runoff model for determining rainfall excess and a unit hydrograph model. An event model simulates the runoff from a one-time rainfall event, whereas a continuous model simulates a time series of daily flows and hydrographs. The curve number model was modified somewhat to include a constant abstraction rate which allows for infiltration during lulls and after the cessation of rainfall but before runoff was ceased. This introduced a new parameter, PHI, which is analogous to the soil saturated hydraulic conductivity, but which is determined solely through optimization studies. The unit hydrograph shape is described with two triangles, the so-called double triangle unit hydrograph, and requires four parameters, the peak flowrate and time to peak of the first triangle, the time base for both hydrographs, and the time to peak of the second triangle. A fifth model parameter, the peak ordinate of the second triangle, is determined from the constraint that the volume under the unit hydrograph equal one basin inch or mm of runoff.

The TVA developed regional prediction equations for each of the model parameters using data from over 500 events on 38 rural, urban and surface mined watersheds in the Tennessee Valley region. Using these equations, the model can be applied to other watersheds within the same physiographic regions with reasonable success, as demonstrated in verification studies by Betson, et al. (1981). However, this model should not be used outside the limits of its regionalization. This was demonstrated by Meadows, et al. (1983) in a study of the application of four unit hydrograph models to watersheds in 14 physiographic provinces across the United States. The results which they obtained for each model basically were acceptable only within the regions of their development.

Most deterministic models are formulated with the kinematic runoff model, of which there are several, including the EPA Storm Water Management Model (Metcalf and Eddy, et al., 1971). the USGS Distributed Rainfall-Runoff Routing Model (Dawdy, et al., 1978), and WITWAT (Green, 1984), to name a few. These models differ, but each is formulated with a 2 or 3 parameter model for infiltration, and kinematic overland and channel routing. The infiltration model parameters generally can be estimated from site measures or as typical values in textbooks and published reports. Similarly, the routing parameters, e.g. Manning's n-value, can be estimated from published sources. Thus, these models are applicable to an ungauged site because model parameters generally are measurable or typical values are known. Confidence in model simulation is high, but should be confirmed through verification studies once local data became available.

The best of both worlds is illustrated by the USGS model. It can be applied directly as a deterministic model, or if local calibration data are available, the soil-moisture accounting and infiltration parameters can be optimized. The USGS terms this version of the model a parametric-deterministic runoff model (Alley, et al., 1980).

Engineers have designed drainage systems for decades using the well-known Rational Method, and have simulated watershed runoff with unit hydrograph models, e.g. SCS curvilinear unit hydrograph. Why is it necessary, or even useful, to work with kinematic stormwater models now? The answer to this question lies in an examination of what kinematic models will do for the engineer – and perhaps it also lies in what the other methods will not do.

First, the role of models in general should be acknowledged. Engineering design of drainage systems and environmental impact assessment of land use change require information about watershed response to prescribed "design" events which most often are extreme events. Since most small basins are not gauged for both rainfall and streamflow, little hydrologic data is available to quantify the necessary response characteristics. Further, if a watershed is gauged, it is unlikely that a suitable "design" event is contained in the record unless the gauge has been in operation for many years. Even so, the data are for the watershed response in its current land use condition and are not a true measure of the watershed response following land use change. To properly quantify the watershed response for "after" development conditions, the record would have to be extended for several years to insure the probability of an adequate number of acceptable events. But the land use change must still be planned, the associated drainage system designed, and impact statements prepared. We do not have the luxury of being able to wait for the data to be collected, so we must resort to prediction methods.

It is widely accepted that mathematical stormwater models are the only available means of making reliable predictions of watershed response to design events and of the effects of land use change on stormwater runoff and quality. It must be stated emphatically, however, that models are not a substitute for field gathered data or knowledge of the hydrologic/hydraulic and water quality processes on the part of the user. No model can predit how a natural system behaves as dependably as direct measurements of the system itself. The principal use of models is in situations where direct measurements are either impossible or impractical, such as the "after" development conditions. When a drainage system is under design, for example, a model will let the designer look at many alternative configurations. More importantly, the designer can answer the "what if" questions, and can do so within a reasonable framework of time and costs. Models also permit a more accurate analysis of complex watershed and drainage systems. The advent of models has changed the engineer from a cookbook artist who relied heavily on judgement to a serious analyst and planner.

The selection of a model typically is a statement of user confidence, which has been defined as "the belief in the reliability or credibility of the results and exists either consciously or subconsciously in the minds of the model user or clientele" (ASCE, 1983). This belief is derived from experiences in the use, development, or testing of a model, from user understanding of watershed hydrologic processes and model representation of these processes, and from confidence in authority, e.g., textbooks, technical journals, and federal agency endorsement. Ultimately, confidence is founded on verification studies at the watershed where the simulations are required.

The keyword is "reliable". When using a model, one must remember the model is merely a mathematical expression of the true system and cannot account for all the subtleties of the various phenomena (processes) involved. Reliable results are those on which the model user can foster the belief that if such an event occurs, the probably runoff hydrograph will be very much like the model predictions.

So why use kinematic models? Perhaps the best answer is that they are deterministic, distributed parameter models that can account for the spatial watershed and rainfall variations and the nonlinearity of the runoff process. In other words, kinematic models are a better model of the true process. Because kinematic models are based on the physics of the runoff process; the model structure is rational, the parameters are measurable or are available from published studies and textbooks, and the model can be applied with a minimum of calibration data. (They can be applied in the absence of calibration data; user confidence is supported by the extensive testing and documentation of kinematic models.) Though young in evolution, there are now several models available for computer use, even personal computer use. Thus, kinematic models are as readily used as other models and have the advantages offered by deterministic models.

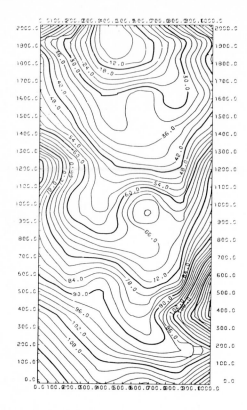

Fig. 10.2   Contour plot of topography

TWO - DIMENSIONAL OVERLAND FLOW MODELLING

Topography and catchment surface characteristics can not be properly accounted for in one–dimensional models in all cases. The cone shaped catchment is a typical example. Also the effect of varying surface roughness, slope and losses is often two–dimensional. Storm patterns cannot be accounted for properly and the assumption of a rectangular hyetograph over the entire catchment is often dangerous. It may be necessary in the case of complex catchments or rain to resort to two–dimensional modelling.

Two–dimensional kinematic equations

One - dimensional equations can be extended into two dimensions as follows:

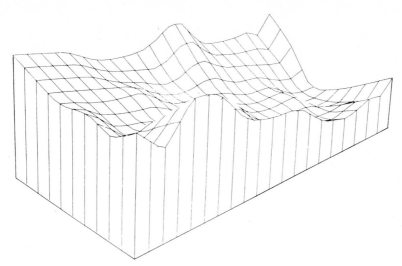

Fig. 10.3   Topography in 3-Dimensions

The continuity equation becomes

$$\frac{\partial y}{\partial t} + \frac{\partial q_x}{\partial x} + \frac{\partial q_z}{\partial z} = i_e \qquad (10.1)$$

where q_x is the flow in the x direction (m^2/s) and q_z is the flow in the z direction (m^2/s).

A proof of this equation can be found in Dronkers (1964). For two dimensional flow two motion equations are required. In kinematic theory these are obtained by assuming

$$S_{ox} = S_{fx} \qquad (10.2)$$

$$S_{oz} = S_{fz} \qquad (10.3)$$

where S_{ox} is the bed slope in the x direction, S_{oz} is the bed slope in the z direction, S_{fx} is the friction slope in the x direction and S_{fz} is the friction slope in the z direction.

For the general form of headloss equation one can obtain

$$q_x = \frac{1}{q_t}(\alpha_x y^m)^2 \qquad (10.4)$$

$$q_z = \frac{1}{q_t}(\alpha_z y^m)^2 \qquad (10.5)$$

where $q_t = (q_x^2 + q_z^2)^{\frac{1}{2}} \qquad (10.6)$

and α_x = function of S_{ox}

α_z = function of S_{oz}

This idea for two dimensional flow was used by Orlob (1972). It will be noticed that q_t is always positive while q_x and q_z can be positive or negative as $(\alpha_x)^2$ and $(\alpha_z)^2$ are functions of S_{ox} and S_{oz} respectively.

Boundary conditions

There are two boundary conditions that can be used on watersheds. One can assume that the water depth at the boundary is always zero and that all the water entering the origin leaves it in the form of a discharge. This has been assumed in all existing theories. One must then define

$$\left[\frac{\partial q}{\partial x}\right]_1^k = i_e \tag{10.7}$$

$$y_1^k = 0.0 \tag{10.8}$$

so $\quad q_1^k = i_e \, \Delta x/2 \tag{10.9}$

One could alternatively assume that the discharge at the origin is controlled by the depth of water at the origin as assumed for the rest of the points. For this case we must then use the same equations as with the other points. The effect of using the two different boundary conditions will be shown later.

Initial conditions

After the first time step it may be assumed that the water depth at all points, except at the origin in the case of the first boundary condition, for the case of an initially dry catchment

$$y_i^1 = i_e \, \Delta t \tag{10.10}$$

The proposed equations may be solved at grid points over a plane provided runoff is adequately described by the kinematic equations. Where there are flow concentrations such as an inland depression, storage will not be accounted for except with a separate routine to account for net volume stored. If outflow eventually occurs when the depression is filled, again a separate routine is needed to detect this.

The effect of channelization, for example rills and furrows in which runoff collects, can be accounted for by reducing the effective dx or dy over which runoff occurs. Where channel side friction is applicable however conduit equations may be required.

The effect of spatially varying soil types and cover can also affect losses to a significant extent. Infiltration, and possible re-emergence of interface flow can be accounted for with a two-layer model with permeable interface. A sample of such a model is discussed later.

Fig. 10.4   Flow directions θ from model

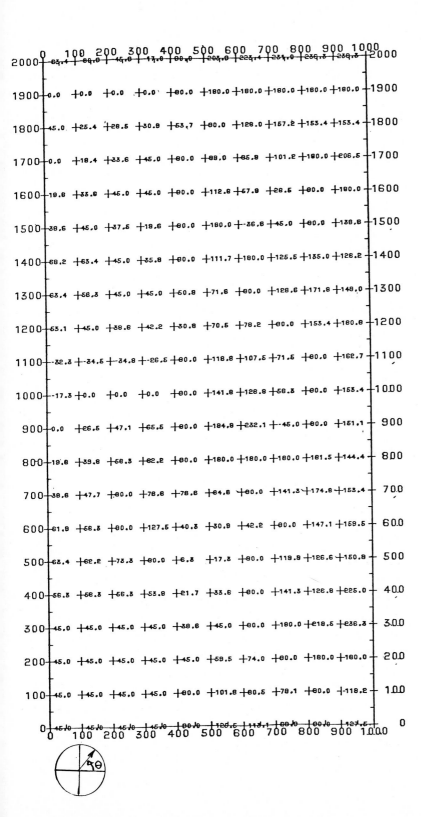

REFERENCES

Alley, W.M., Dawdy, D.R. and Schaake, J.C., Jr., 1980. Parametric-deterministic urban watershed model. J. Hydr. Div. ASCE, Vol. 106, No. HY5, pp. 679-690.

ASCE, 1983. Quantification of land use change effects upon hydrology, by the Task Committee on Quantifying Land Use Change Effects, R.P. Betson, Chmn., presented at the July 20-22, 1983. ASCE Irrigation and Drainage Division Specialty Conference, held at Jackson, Wyoming.

Betson, R.P., Bales, J. and Pratt, H.E., 1980. Users Guide to TVA-HYSIM, A hydrologic program for quantifying land use change effects. EPA-600/7-80-048, Tennessee Valley Authority, Water Systems Development Branch, Norris, Tennessee.

Betson, R.P., Bales, J. and Deane, C.H., 1981. Methodologies for assessing surface mining impacts. Report No. WR28-1-550-108, Tennessee Valley Authority, Water Systems Development Branch, Norris, Tennessee.

Dawdy, D.R., Schaake, J.C., Jr. and Alley, W.M., 1978. User's guide for distributed routing rainfall-runoff model. U.S. Geological Survey Water Resources Investigations 78-90.

Dooge, J.C.I., 1973. Linear theory of hydrologic systems. U.S. Dept. of Agriculture, Agricultural Research Service, Tech. Bull. No. 1468.

Dronkers, J.J., 1964. Tidal computations in rivers and coastal waters. North Holland Publishing Co., Amsterdram.

Green, I.R.A., 1984. WITWAT stormwater drainage program – Theory, Applications and User's Manual. Report No. 1/1984, Water Systems Research Programme, Dept. of Civil Engineering, University of the Witwatersrand, Johannesburg, South Africa.

Meadows, M.E., Howard, K.M. and Chestnut, A.L., 1983. Development of models for simulating stormwater runoff from surface coal mined lands: Unit hydrograph models. Report No. G5115213, Vol. 1, U.S. Dept. of the Interior, Office of Surface Mining, Division of Research, Washington, D.C.

Orlob, G.T., 1972. Mathematical modelling of estuarial systems. International Symposium on mathematical modeling techniques in water resources systems, Editor Asit K. Biswas. Proceedings Volume 1.

Metcalf and Eddy, Inc., University of Florida, and Water Resources Engineers, 1971. Storm water management model. U.S. Environmental Protection Agency, Washington, D.C.

Overton, D.E. and Meadows, M.E., 1976. Stormwater Modeling, Academic Press, New York, N.Y.

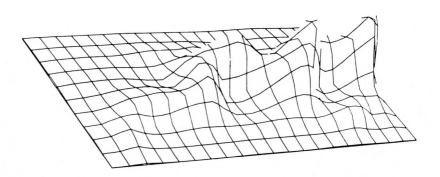

Fig. 10.5   Water depth variation at t = 8 min over the catchment

CHAPTER 11

APPLICATIONS OF KINEMATIC MODELLING

APPROACHES

This chapter contains examples of kinematic stormwater simulation models and their application to rural and urban watersheds. These models were selected from the range of available models because they are simple in concept and structure, have been tested extensively, and are representative of the approaches taken in developing kinematic watershed models. For these reasons, they should help the reader to more fully understand kinematic models and their applications.

The reader is reminded that with any watershed model, approximations and simplifications are made. Previously, we have seen that kinematic models are simplifications to the dynamic wave models; and that their solution, whether analytical or numerical, requires approximations to the watershed geometry, drainage layout, rainfall pattern, etc. The examples in this chapter illustrate different approaches to making these approximations.

A MODEL FOR URBAN WATERSHEDS

A model that has been successfully applied to urban watersheds is the U.S. Geological Survey model, DR3M (Dawdy, et al., 1978). This model combines the soil moisture accounting and rainfall excess components of the model developed by Dawdy and others (1972) with the kinematic wave routing components of the model developed by LeClerc and Schaake (1973). Input to the model includes daily rainfall, storm rainfall, daily pan evaporation and a physical definition of the drainage basin discretized into as many as 50 segments, including overland flow, channel and reservoir segments. During storm days, the model generates a simulated discharge hydrograph based on input data from as many as three rain gauges. The model consists of two main sets of components: parametric rainfall excess and deterministic runoff routing components.

Parametric Rainfall Excess Components

The parametric rainfall excess components are a soil moisture accounting component, an infiltration component, an impervious area rainfall excess component, and an optimization component. A substantial part of the rainfall excess components was adopted from a model developed by Dawdy et al. (1972). This component is used during model calibration to establish optimal parameter values for site infiltration and soil moisture storage functions.

Soil Moisture Accounting

The soil moisture accounting component determines the effect of antecedent conditions on infiltration. Soil moisture is modelled as a two layered system, one representing the antecedent base moisture storage (BMS), and the other, the upper wetted part caused by infiltration into a saturated moisture storage (SMS).

During rainfall days, moisture is added to SMS based on the Philip infiltration equation (Philip, 1954). On other days, a specified proportion of daily rainfall (RR) infiltrates into the soil. Irrigation (for example, lawn watering) can be included in the daily water balance. This is achieved through user supplied irrigation rates for each month. If a daily precipitation is less than the daily irrigation rate, the daily precipitation is set equal to the irrigation rate.

Evapotranspiration takes place from SMS, based on availability, otherwise from BMS, with the rate determined from pan evaporation multiplied by a pan coefficient (EVC). Moisture in SMS drains into BMS with a controlling parameter (DRN) determining the rate. Storage in BMS has a maximum value (BMSN) equivalent to the field capacity moisture storage of an active zone. Zero storage in BMS is assumed to correspond to wilting point conditions in the active soil zone. When storage in BMS exceeds BMSN, the excess is spilled to deeper storage. These spills could be the basis for routing interflow and baseflow components, if desired. However, this option is not included in the present version of the model. A schematic flow chart of the soil moisture accounting component is shown in Figure 11.1.

Infiltration Component

Infiltration is computed with the Philip equation (Philip, 1954), which is merely a variation to the Green and Ampt equation. One form of the Green and Ampt equation is

$$\frac{dF}{dt} = (1 + \frac{H+P}{Z})$$

(11.1)

where F is the accumulated infiltration depth, K is the effective

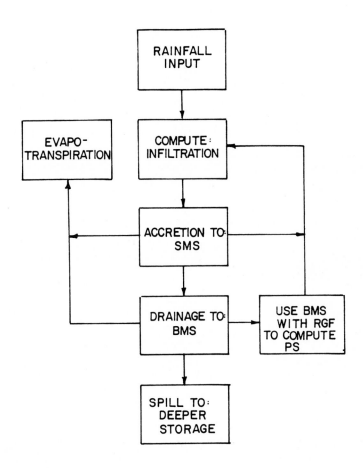

Fig. 11.1   Schematic of DR3M soil moisture accounting component

hydraulic conductivity, H is the depth of water ponded on the soil surface, P is the wetting front section, and Z is the depth to the wetting front. Using the relationship

$$Z = \frac{F}{\theta_s - \theta_i}$$ (11.2)

in which θ_s is the volumetric soil moisture content at saturation and θ_i is the initial (unsaturated) moisture content, Eq. 11.1 is transformed into the Philip equation.

$$\frac{dF}{dt} = K\left[1 + \frac{H + P(\theta_s - \theta_i)}{F}\right]$$ (11.3)

Since the wetting front suction is generally several orders of magnitude greater than the depth of ponded water, the H term may be ignored. The mnemonic identifiers used to designate the resulting infiltration are

$$FR = KSAT\ (1 + \frac{PS}{SMS})$$ (11.4)

in which FR=dF/dt, KSAT=K, PS=P($\theta_s - \theta_i$), and SMS=F.

The wetting front suction is not constant, but varies with the soil moisture condition. The effective value of PS is assumed to vary linearly between a wilting point and field capacity, and is computed with the relationship

$$PS = PSP\left[RGF - (RGF - 1)\frac{RMS}{BMSN}\right]$$ (11.5)

in which BMS is the initial moisture storage in the soil column; BMSN is the moisture storage in the soil column at field capacity; PSP is the effective value of PS at field capacity; and RGF is the ratio of PS at wilting point to that at field capacity. This relationship is shown in Figure 11.2.

Point potential infiltration (FR) computed by the Philip equation is converted to effective infiltration over the basin using the scheme of Crawford and Linsley (1966). Letting SR represent the supply rate of rainfall for infiltration and OR represent the rate of generation of rainfall excess, the equations are

$$OR = \frac{SR}{2FR};\quad \text{if } SR < FR$$ (11.6a)

$$OR = SR - \frac{FR}{2};\quad \text{if } SR > FR$$ (11.6b)

A schematic of these relationships is shown in Figure 11.3. The rainfall excess rate, OR, is represented by the area between the dashed SR line and the linear infiltration capacity curve. The parameters for soil moisture accounting and infiltration are enumerated in the following:

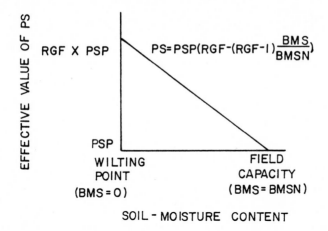

Fig. 11.2   Relationship   determining   effective   value   of   soil-moisture
potential (PS)

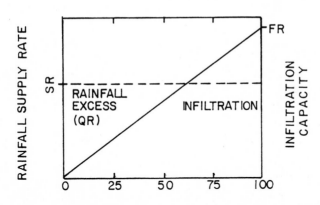

Fig. 11.3   Relationship   determining   rainfall   excess   (OR)   as   function
of   maximum   infiltration   capacity   (FR)   and   supply   rate   of
rainfall (SR)

1. Soil Moisture Accounting. The parameters consist of: (a) DRN – A constant drainage rate for redistribution of soil moisture between SMS and BMS, in inches per day; (b) EVC – A pan coefficient for converting measured pan evaporation to potential evapotranspiration; (c) RR – The average proportion of daily rainfall that infiltrates into the soil for the period of simulation excluding storm days; and (d) BMSN – Soil moisture storage at field capacity, in inches.

2. Infiltration. The parameters consist of: (a) KSAT – The hydraulic conductivity at natural saturation, in inches per hour; (b) RGF – Ratio of suction at the wetting front for soil moisture at wilting point to that at field capacity.

Impervious Area Rainfall Excess Component

Two types of impervious surfaces are considered by the model. The first type, effective impervious surfaces, are those impervious areas that are directly connected to the channel drainage system. Roofs that drain onto driveways, streets and paved parking lots that drain onto streets are examples of effective impervious surfaces. The second type, non-effective impervious surfaces, are those impervious areas that drain to pervious areas. An example of a noneffective impervious area is a roof that drains onto a lawn.

The only abstraction from rainfall on effective impervious areas is impervious retention. This retention, which is user specified, must be filled before runoff from effective impervious areas can occur. Evaporation occurs from impervious retention during periods of no rainfall.

Rain falling on noneffective impervious areas is assumed to runoff onto the surrounding pervious area. The model assumes this occurs instantaneously and that the volume of runoff is uniformly distributed over the contributing pervious area. This volume is added to the rain falling on the pervious areas prior to computation of pervious area rainfall excess.

Optimization Component

An option is included in the model to calibrate the soil moisture and infiltration parameters for drainage basins having measured rainfall runoff data. The method of determining optimum parameter values is based on an optimization technique devised by Rosenbrock (1960).

Impervious area is not included as a parameter to be optimized, but is a parameter to which simulated runoff volumes are very sensitive. Therefore, values of imperviousness should be determined accurately before using the optimization option. If initial estimates of imperviousness are grossly in error, resulting volumes and peaks will be grossly in error. In that case, estimates of imperviousness must be adjusted by the modeler. This adjustment may include revising estimates of the effective and noneffective impervious areas, perhaps by trial and error.

Deterministic Runoff Routing Components

After determining "optimum" parameter values and computing the time series of rainfall excess, control in the model is transferred to the runoff routing component. The mathematical representation of an urban basin requires discretization of the total drainage area into a set of segments. There are three basic types of segments defined for the model: channel segments, overland flow segments and reservoir segments. There is wide flexibility to the approach one can take in dividing a basin into segments for runoff computations. Guidelines for basin segmentation are presented by Alley and Veenhuis (1979) and Dawdy, et al., (1978).

Channel and Overland Flow Segments

A channel segment is permitted to receive upstream inflow from as many as three other segments, including other channel segments and reservoir segments. It also may receive lateral inflow from overland flow segments. The overland flow segments receive uniformly distributed lateral inflow from rainfall excess. A schematic illustrating the relationships between channel and overland flow segments is shown in Figure 11.4 - 5.

Kinematic wave theory is applied in the rainfall runoff model to both overland flow and channel routing. The necessary equations to be solved for each channel and overland flow segment are

$$\frac{\partial Q}{\partial x} + \frac{\partial A}{\partial t} = q_i \tag{11.7}$$

$$Q = aA^b \tag{11.8}$$

in which the terms are as previously defined.

Finite difference approximations are used to solve Eqs. 11.7 and 11.8. To avoid the convergence and stability problems that can occur with particular numerical grid spacings (i.e. the relative sizes of Δt and Δx), two finite difference methods of solution are used to solve for

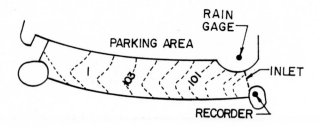

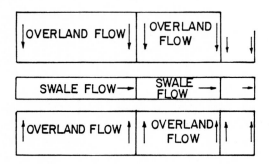

Segment	Length (ft)	Slope (ft/ft)	Inflow to segment	
			Lateral inflow	Upstream inflow
①	36	.019	Rainfall	———
②	20	.0167	Rainfall	———
③	25	.019	Rainfall	———
④	165	.0148	① and ②	———
⑤	100	.0213	① and ②	④
⑥	50	.0213	② and ③	⑤

PHYSICAL CHARACTERISTICS OF COMPONENTS
IN THE SCHEMATIC REPRESENTATION

Fig. 11.5   Discretization of urban catchment into segments

Q and A at the unknown grid points. The chosen solution procedure is made in the model program and depends upon the ratio, θ, of the kinematic wave speed to $\Delta x / \Delta t$.

$$\theta = b \frac{\Delta t}{\Delta x} \frac{Q_3}{A_3} = abA^{b-1} \frac{\Delta t}{\Delta x} \qquad (11.9)$$

in which Q_3 is the discharge at node point 3 in the finite difference grid as shown in Figure 11.6. If θ is greater than or equal to unity, the equations used are

$$Q_4 = Q_2 + q\Delta x - \frac{\Delta x}{\Delta t} (A_2 - A_1) \qquad (11.10)$$

$$A_4 = (\frac{Q_4}{a})^{1/b} \qquad (11.11)$$

This involves only mesh points 1, 2 and 4. If θ is less than unity, the equations used are

$$A_4 = A_3 + q t + \frac{\Delta t}{\Delta x} (Q_1 - Q_3) \qquad (11.12)$$

$$Q_4 = aA_4^b \qquad (11.13)$$

Δx and Δt values are chosen to ensure about 10 nonzero ordinates under the rising limb of an equilibrium hydrograph and to keep computational errors within acceptable bounds. The U.S.G.S. recommends that Δt be selected as

$$\Delta t = 0.1 (t_{eo} + t_{ec}) \qquad (11.14)$$

where t_{eo} and t_{ec} are the kinematic overland and channel times of equilibrium, respectively. Computational error should be minimized if Δx and Δt are selected so that the characteristic passing through point 1 also passes through point 4. Accordingly, it is recommended that Δx be selected as

$$\Delta x_o = \frac{L_o}{t_{eo}} \Delta t \qquad (11.15a)$$

$$\Delta x_c = \frac{L_c}{t_{ec}} \Delta t \qquad (11.15b)$$

for the overland and channel segments, respectively. When Eqs. 11.14 and 11.15 result in non-integer values, the user must round to the nearest integer.

Reservoir Segments

Provision is made in the model for reservoir routing based on the continuity equation. Either of two routing methods can be used. One method is linear storage routing

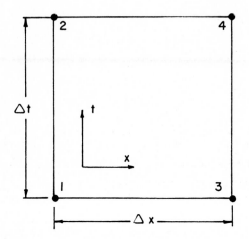

Fig. 11.6   Four point finite difference grid

S = CO $\qquad$ (11.16)

in which S is the storage; Q is the outflow; and C is a constant. Alternatively, the modified Puls routing method can be used

$$\frac{2S_2}{\Delta t} + O_2 = I_1 + I_2 + \frac{2S_1}{\Delta t} - O \qquad (11.17)$$

in which I is the inflow to the reservoir and the subscripts 1 and 2 refer to the beginning and end of the time interval, Δt, respectively. The modified Puls method utilizes a table of storage outflow values as supplied by the user.

There are many ways of accounting for storage with the kinematic method, all of which are, due to the limitation of the kinematic method, approximations. That is, due to the fact that the dynamic equation omits acceleration and deceleration of the water in time and space, wedge storage is omitted. Storage can only be included as channel type storage (see Chapter 9) if the discharge relationship can be described in terms of a kinematic type of discharge – depth equation.

The continuity equation in the kinematic equations can however be used to account for the lag effect of storage in one of two ways. Inflow to a reach can be spread over the full surface area of the reach as demonstrated in the example later in this chapter, or the stream (or overland flow) surface area can be replaced by a storage area at the junctions of reaches with conduit reaches as indicated in the program in Chapter 8.

Example Application

The model was applied to the Sand Creek Tributary watershed near Denver, Colorado. This drainage basin is a 183 acre area of predominantly single family residential land use with some multifamily land use, a church, a recreational center, a fire station, and two small parks. The basin has some storm sewers in its upper end but relies mostly on street gutters and concrete lined open ditches for flow conveyance. Detailed records of rainfall and streamflow are collected at 5 minute intervals. A stage discharge relation was developed using flow profile analysis and discharge measurements made during storm runoff.

Two sample runs are discussed. The first run was an optimization run to calibrate the model on an antecedent period of record. In the second run, the soil moisture accounting and infiltration parameters were set to their final values from the first run and ten storm events were simulated.

Before any simulations were performed, the watershed was delineated into sub-basins (overland flow segments) and a drainage network (channel segments). A schematic showing how the watershed was approximated with the overland and channel segments is given in Figure 11.7.

The rationale behind the basin segmentation is as follows: starting at the basin outfall, it was first noted that the major drainage system of the basin consisted of concrete lined ditches which were located in the positions marked by channel segments CH20, CH21 and CH22. In analyzing the reach of concrete lined ditch comprised of CH20 and CH21, it was noted that a street, which drained 14 acres of land ØF03, intersected this reach. Therefore, this reach of concrete lined ditch was subdivided into channel segments CH20 and CH21, and the intersecting street was designated as channel segment CH23. Overland flow segments ØF01,

220

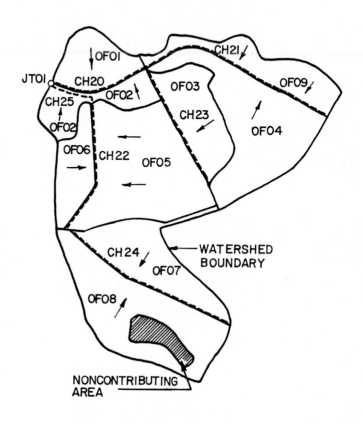

OFO2   OVERLAND FLOW SEGMENT AND NUMBE[R]

CH23   CHANNEL SEGMENT

JTOI   JUNCTION AND NUMBER

⟶   GENERAL DIRECTION OF OVERLAND FLO[W]

Fig. 11.7 Schematic representation of Sand Creek tributary watershed

ØF02, ØF03, ØF04 and ØF09 were then delineated based on this channel segmentation. It should be noted that overland flow segment F04 does not have balanced lengths of overland flow to CH21. To further subdivide this overland flow segment would also require that segments CH21 and ØF09 be further subdivided.

The unallocated concrete lined ditch was then assigned as channel segment CH22. Overland flow segments ØF06 and ØF05 were then delineated. To avoid the need to subdivide channel segment CH20 which would require subdividing overland flow segments ØF01 and ØF02, channel segment CH25 was used to bypass channel segment CH20. A junction segment, JT01, was required to sum the flow from the two channel segments at the outlet of the basin. Finally, the remaining part of the basin was drained by a street which was assigned as channel segment CH24.

Once the basin was segmented, the sub-basin boundaries were field checked and representative channel cross sections determined. Channel slopes were determined from the drainage maps, and overland flow slopes were estimated from the U.S. Geological Survey topographic map for the area and the street corner elevations shown on the City of Denver drainage maps. Sub-basin areas were planimetered and lengths of over-land flow were computed by dividing the area of each sub-basin, in square feet, by the length, in feet of the channel segment into which it contributes lateral inflow.

TABLE 11.1 Model Simulation Results for Sand Creek Tributary Watershed

Runoff Event Number	Date	Runoff Volume in inches		Peak Flow in cfs	
		Measured	Simulated	Measured	Simulated
1	7-12-73	0.08	0.08	32	23
2	7-19-73	0.16	0.19	68	74
3	7-22-73	0.055	0.052	22	14
4	7-24-73	0.33	0.28	104	97
5	7-30-73	0.063	0.082	32	28
6	8-07-73	0.70	0.76	236	280
7	9-11-73	0.073	0.14	48	58
8	9-11-73	0.23	0.16	143	68
9	7-22-74	0.20	0.32	98	117
10	7-30-74	0.53	0.47	251	216

The period of record to be simulated was July 12, 1973 to July, 30, 1974. To establish initial moisture conditions for the beginning data, an optimization run was conducted for the period May 1, 1973 to July 12, 1973. Using input values for rainfall daily pan evaporation and recorded runoff the model was "calibrated" by determining optimal values for the infiltration and soil moisture accounting parameters. A second run was then conducted to simulate watershed runoff (hydrographs) during the specified simulation period. Results are shown in Table 11.1.

A MODEL FOR RURAL WATERSHEDS

The model described for urban stormwater simulation could be applied easily to rural watersheds. In this section, however, let us examine another model which has been applied only to rural watersheds, and which uses finite elements to solve the kinematic equation instead of finite differencing. The following model was developed for rural water-sheds in agricultural land use, but it has been suggested it also can be applied to surface mining disturbed watersheds.

A finite element storm hydrograph model (FESHM) has been developed at Virginia Polytechnic Institute and State University as part of a program to develop a distributed parameter model to simulate flow on ungauged watersheds (Ross, et al., 1978 and 1982). Spatial variability was a requirement because a long range goal is to be able to simulate not only runoff from mixed land use watersheds, but also waterborne pollutants. Thus, FESHM was developed to integrate spatial and temporal variations in climatic and watershed characteristics.

The model consists of two major components: a precipitation excess generator and a flood routing algorithm which routes the excess along overland flow elements and down the stream channel elements.

Precipitation Excess

The calculation of rainfall excess depends on the spatial distribution of two watershed characteristics, land use and soil mapping units. A map of land use patterns is superimposed on the watershed site map, defining soils to create a hydrologic response unit (HRU) map. Each area with a unique land use and soil mapping combination is referred to as an HRU. A given rainfall on the watershed will result in a different amount of precipitation excess from each HRU.

The amount of precipitation excess is determined using Holtan's infiltration equation (Holtan, 1961). This infiltration model has been applied to a wide range of data by Shanholtz and Lillard (1970) and Holtan, et al., (1975) with reasonable success. It was included in FESHM primarily because the data necessary to define model parameters closely parallel the concepts for dividing a drainage area into HRUs.

Flow Routing

The second part of FESHM routes precipitation excess to the outlet of the watershed. To accomplish this, the watershed is divided into overland and streamflow elements as shown in Figure 11.8. The number of elements to be used depends on the hydraulic and hydrologic heterogeneity of the watershed. The HRUs that occur in each overland flow element are catalogued, and the rainfall excess from each HRU is weighted by its fractional area in the element. This rainfall excess is then routed through overland flow elements using a finite element approximation of the kinematic wave model.

Using the Galerkin technique (Lapidus and Pinder, 1982) and linear variation of parameters within an element, the element equation becomes

$$\ell \begin{bmatrix} 1/3 & 1/6 \\ 1/6 & 1/3 \end{bmatrix} \begin{Bmatrix} \dot{A}_1 \\ \dot{A}_2 \end{Bmatrix} + \begin{bmatrix} -1/2 & 1/2 \\ -1/2 & 1/2 \end{bmatrix} \begin{Bmatrix} Q_1 \\ Q_2 \end{Bmatrix} - \ell q \begin{Bmatrix} 1/2 \\ 1/2 \end{Bmatrix} = 0 \tag{11.18}$$

where script "ℓ" is the element length, equivalent to the flow length across each element. The time differential of area is replaced by a simple finite difference approximation. Thus

$$\frac{\partial A}{\partial t} = \frac{A(t+\Delta t) - A(t)}{\Delta t} \tag{11.19}$$

The final element equation is

$$\ell/\Delta t \begin{bmatrix} 1/3 & 1/6 \\ 1/6 & 1/3 \end{bmatrix} \begin{Bmatrix} A_1 \\ A_2 \end{Bmatrix}_{t+\Delta t} - \ell/\Delta t \begin{bmatrix} 1/3 & 1/6 \\ 1/6 & 1/3 \end{bmatrix} \begin{Bmatrix} A_1 \\ A_2 \end{Bmatrix}_t \tag{11.20}$$

$$+ \begin{bmatrix} -1/2 & 1/2 \\ -1/2 & 1/2 \end{bmatrix} \begin{Bmatrix} Q_1 \\ Q_2 \end{Bmatrix}_t - \ell q \begin{Bmatrix} 1/2 \\ 1/2 \end{Bmatrix} = 0$$

Model Application

FESHM has been tested on watersheds in seven states covering a wide range of land use, topography and climatic conditions. Drainage areas ranged from approximately 2 acres to 193 square miles. The size

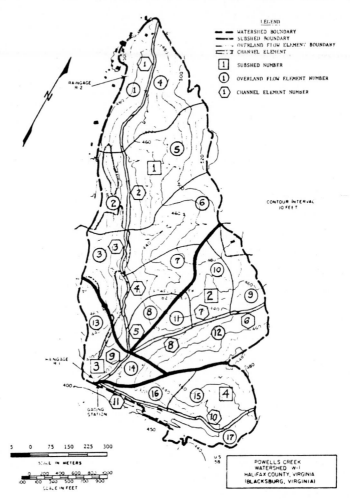

Fig. 11.8   Watershed map showing overland and channel finite elements

of the watershed did not limit its application; the accuracy of simulations, however, most likely was a function of the quality and resolution of the available data.

The model was applied to six watersheds in Virginia, ranging in size from 183 to 1,058 acres (1 acre = 0.4047 hectares). Sixteen storm events on these watersheds were simulated and the simulated hydrographs compared with the observed. These runs were made without any efforts to calibrate the model, and therefore are typical of the model's performance on ungauged watersheds. The results are summarized in Table 11.2. The mean error in the predicted storm volume was 4.4 percent, with a standard deviation of 49.9 percent. The mean error in the predicted peak discharge was 22.6 percent, with a standard deviation of 50.1 percent. The large standard deviation is due largely to the inclusion of some very small (volume) storms. The model tends to simulate best those storms with a return period of 20 years or greater.

An example of the simulation results of a storm event on Powell's Creek is shown in Figure 11.9. The HRU map is given in Figure 11.10, and the subdivision of the watershed into eight finite elements is shown in Figure 11.11. For this storm, the error in peak discharge was 3.4 percent, and the error in storm volume was 2.1 percent. A sensitivity analysis of element size indicated little improvement to using smaller elements.

TABLE 11.2   Comparison of Runoff Volume and Peak Flow for Simulated and Recorded Flows in Virginia Watersheds (Ross, 1978)

Watershed	Storm Event	Runoff Volume (in) Recorded	Simulated	Peak Flow (cfs) Recorded	Simulated
Powells Creek	10/10/59	0.73	0.46	109.88	114.34
	5/31/62	0.92	1.34	241.10	359.71
	7/11/65	2.06	2.38	419.82	775.13
Pony Mountain Branch	6/12/58	0.44	0.46	83.29	78.32
	6/24/58	0.43	0.44	91.26	94.44
	9/19/60	0.73	0.47	45.74	59.95
Rocky Run Branch	7/23/70	1.44	1.13	327.80	544.32
	10/ 5/72	5.79	3.26	609.77	551.00
Crab Creek	8/21/66	0.19	0.26	179.20	231.16
	10/24/71	0.57	0.30	180.54	137.59
	6/16/76	0.49	0.31	205.58	148.34
Brush Creek	7/22/59	0.44	1.08	791.47	1,741.56
	9/30/59	1.14	0.86	924.70	434.69
	5/28/73	0.27	0.43	296.38	623.58
Chestnut Branch	8/23/67	0.67	0.63	416.08	595.84
	8/ 4/74	0.43	0.49	339.65	448.43

226

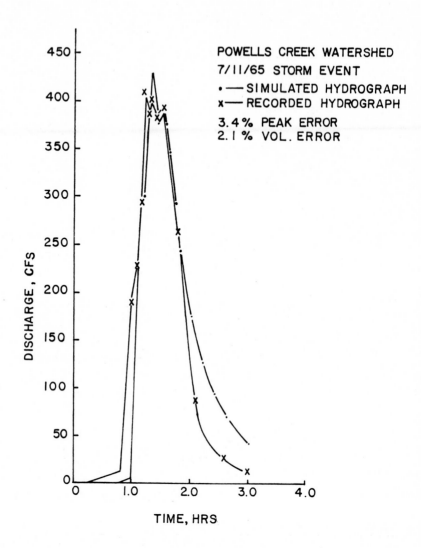

Fig. 11.9   Comparison of simulated and recorded hydrographs Powells
              Creek watershed, Virginia

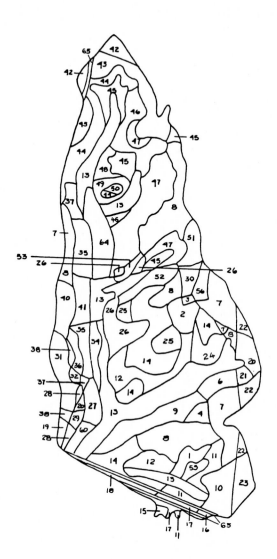

Fig. 11.10 HRU map of Powells Creek watershed

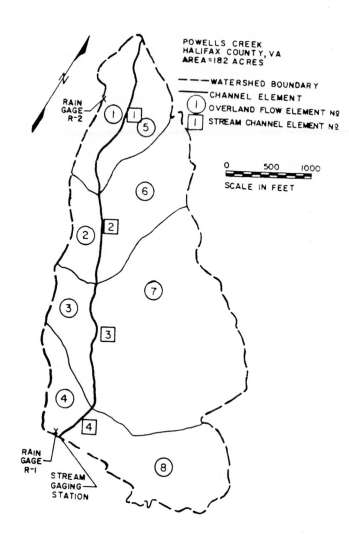

Fig. 11.11 Finite element map for Powells Creek watershed

OVERLAND FLOW AND STREAMFLOW PROGRAM

The simplicity with which a computer program can be assembled from
the basic kinematic equations is demonstrated here. The program written
in BASIC for an HP85 micro computer is appended. It is a simplistic
program incorporating overland flow and channel flow in series. Con-
secutive channels can feed into designated downstream channels, enabling
branches to be included.

The channels are assumed for simplicity to be rectangular in cross
section with flood planes on either side (Fig. 11.12), which in fact can
form the overland flow planes. Flow planes are assumed to be rectangular
and of uniform slope and roughness.

An assembly of planes and channels of the type envisaged is
illustrated in Fig. 11.13.

The model was developed to study the effect of bank storage on flood
routing. The flood planes act as dead storage – flow on them is assumed
lateral – from or to the channel where the longitudinal flow occurs.
Channel depressions can also be included – by widening the stream bed.
Flow is obviously assumed to be kinematic – that is backwater effects
and unsteady flow are neglected.

Overland flow of sub-catchments is calculated using the kinematic
equations for rectangular planes. Inflow is the net rainfall and outflow
is assumed over the full width B of the plane. Flow rate is calculated
from an equation of the form:

$$Q = B \alpha y^m \tag{11.21}$$

where $\alpha = S^{1/2}/n$

S = slope in direction of flow

n = Manning roughness coefficient

$m = 5/3$

Inflow to channel reaches can be from both sub-catchments and
upstream channels, but not rain, as the width is assumed negligible,
as well as losses along the channel.

All storage in channels and overland is assumed to be prism storage
i.e. no wedge storage or difference between bed slope and water surface
slope is permitted. This is in accordance with the kinematic simplification
but provided distance intervals are limited, it is not inaccurate. Flow
off the sub-catchments is also assumed to be a function of the average
depth at the outlet end, so if the variation in depth is likely to be
significant a cascade of planes or planes leading into wide channels
representing planes may be preferable.

The kinematic equations are able to accommodate storage by re-writing as follows the continuity equation

$$\frac{\partial q}{\partial x} + \frac{\partial y}{\partial t} = i_e \qquad (11.22)$$

$$LBdy = dt(Ai_e + Q_i - Q_o) \qquad (11.23)$$

where A is the area of the sub-catchment with excess rainfall rate i_e and $Q_i - Q_o$ is the upstream inflow minus downstream outflow which in turn are functions of the water depth y and LB is the surface length times width of the channel plus flood planes (based on the water depth at the previous time interval as a simple explicit solution is used). In fact the term Ai_e falls away for the channel storage computations and LB is equal to A for overland flow computations.

Since the simulation is not strictly a finite difference solution to the differential continuity equation but merely a flow balance at successive nodes or junctions the speed of propagation of disturbances is not strictly corrrect. Thus some numerical diffusion is bound to occur unless all reach lengths are proportional to dx/dt, the wave speed. The advantages of a variable reach length, however, appear to outweigh the disadvant-ages. That is data can be fed in in natural (unequal) channel lengths, and time intervals can be extended above what is normally required for kinematic simulation.

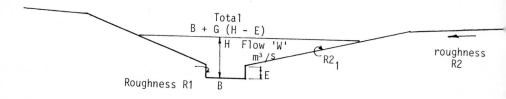

Fig. 11.12 Channel section

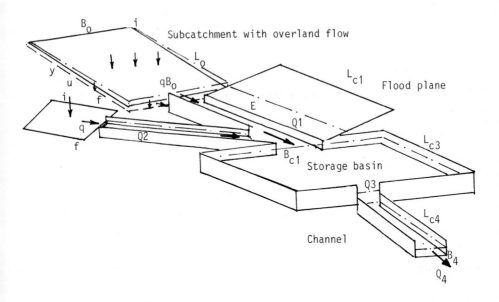

B_0 i Subcatchment with overland flow
L_0
y
u qB_0
i f E L_{c1} Flood plane
q Q1
f Q2
B_{c1} Storage basin L_{c3}
Q3
Channel L_{c4}
B_4
Q_4

Fig. 11.13   Possible layout of sub-catchments and channels for kinematic flood plane model

Data input

The basic program requests interactively the following data. The data for each line is typed in consecutively with commas separating the numbers.

First Input Line

Simulation duration, minutes

Interval between successive flow calculations, minutes

Interval between tabulation of flows and depths for each channel, minutes.

Second Input Line

Preceding week's rainfall in mm

Channel no. at which a hydrograph is required

Estimate of maximum flow in channel (m^3/s) for hydrograph plot axis scaling.

Third Series of Input (each Time interval)

Rainfall intensity in mm/h (assumed uniformly distributed)

Data Lines at end of program (stored as a file to save input each run).

Downstream channel no. into which the sub-catchment discharges

Sub-catchment surface area, m^2

Overland and streamflow program listing

```
  5 PRINT
 10 PRINT "OVERLAND & CHANNEL KI
    NEMATIC  FLOW SIMULATION CAT
    SIM"
 20 DIM J(30),A(30),L(30),S(30),
    F(30),U(30),N(30),O(30),X(30
    ),Z(30),B(30),V(30),R(30)
 30 DIM G(30),E(30),H(30),W(30),
    Q(30),Y(30),P(30,6),C(30,6)
 40 DISP "SIMULATIONt,INTERVAL,T
    ABULATIONt(all mins)";
 50 INPUT T1,T2,T4
 52 DISP "PRECEEDING WK RAINmm,H
    YDROGPH CHANL no,QMAXm3/s";
 53 INPUT R5,K5,Q1
 55 PRINT
 60 PRINT "CHANL Qm3s DEPTHm"
 70 I1=0
 80 A1=0
 85 M=5/3
 90 FOR I=1 TO 50
100 READ J(I),A(I),L(I),S(I),R3,
    F(I)
101 IF I>1 THEN 105
105 IF J(I)=0 THEN 200
110 S(I)=S(I)^.5/R3
120 Y(I)=0
130 Q(I)=0
140 A1=A1+A(I)
150 I1=I1+1
160 NEXT I
200 K1=0
210 FOR K=1 TO 50
220 READ N(K),O(K),X(K),Z(K),R4,
    B(K),G(K),E(K)
222 V(K)=Z(K)^.5/R4
230 Z(K)=Z(K)^.5/R4
240 W(K)=0
241 H(K)=.1
242 C(K,1)=0
243 P(K,1)=0
245 K1=K1+1
250 IF O(K)=0 THEN 280
270 NEXT K
280 FOR K=1 TO K1
282 IF N(K)<>K5 THEN 290
284 K5=k
290 K2=0
300 FOR I=1 TO I1
310 IF J(I)<>N(K) THEN 350
320 K2=K2+1
330 P(K,K2)=I
350 NEXT I
360 P(K,K2+1)=0
365 K4=0
370 FOR K3=1 TO K1
380 IF O(K3)<>N(K) THEN 410
390 K4=K4+1
400 C(K,K4)=K3
410 NEXT K3
```

```
420 C(K,K4+1)=0
430 NEXT K
450 GCLEAR
460 SCALE -T2,T1,-(Q1/20),Q1
470 XAXIS 0,60,0,T1
480 YAXIS 0,100,0,Q1
490 MOVE T2,-(Q1/20)
500 LABEL "Thours"
510 MOVE -T2,Q1*.9
515 Q1=INT(Q1)
520 LABEL "Qm3/s- "&VAL$(Q1)&" I
    N CHANL "&VAL$(K5)
540 FOR T6=0 TO T1-T4 STEP T4
545 FOR T5=T2 TO T4 STEP T2
548 T=T6+T5
550 DISP "RAINmm/h"
555 INPUT R(1)
570 FOR I=1 TO I1
572 R(I)=R(1)
575 U(I)=F(I)*(1+((1000/(R5+10))
    ^.5-1)*2.718^(-(.001*T*60)))
580 Y(I)=(R(I)-U(I))/3600000*T2*
    60+Y(I)-Q(I)/A(I)*T2*60
590 IF Y(I)>0 THEN 620
600 Q(I)=0
610 GOTO 630
620 Q(I)=S(I)*Y(I)^M*A(I)/L(I)
630 NEXT I
640 FOR K=1 TO K1
650 Q1=-W(K)
660 FOR K2=1 TO 6
670 IF P(K,K2)=0 THEN 700
680 Q1=Q1+Q(P(K,K2))
690 NEXT K2
700 FOR K4=1 TO 6
710 IF C(K,K4)=0 THEN 740
720 Q1=Q1+W(C(K,K4))
730 NEXT K4
740 B1=B(K)
742 IF H(K)<=E(K) THEN 750
744 B1=B1+(H(K)-E(K))*G(K)
750 H(K)=H(K)+Q1*T2*60/X(K)/B1
752 IF H(K)>0 THEN 760
754 H(K)=.01
760 NEXT K
770 FOR K=1 TO K1
780 W(K)=Z(K)*H(K)^M*B(K)
782 IF H(K)<=E(K) THEN 800
785 W(K)=W(K)+V(K)*G(K)*(H(K)-E(
    K))^(M+1)/2^M
795 IMAGE 3D,DDDD.DD,DDD.DDD
800 NEXT K
802 Q2=0
803 FOR I=1 TO I1
804 Q2=Q2+A(I)*(R(I)-F(I))/36000
    00
805 NEXT I
807 MOVE T,Q2
808 PLOT T,Q2
810 MOVE T,W(K5)
```

```
820 PLOT T,W(K5)
825 NEXT T5
830 PRINT "TIME mins",T
831 FOR K=1 TO K1
832 PRINT USING 795 ; N(K),W(K),
    H(K)
833 NEXT K
834 NEXT T6
835 COPY
840 END
1010 DATA 1,60000000,10000,.02,.
     1,5
1020 DATA 2,60000000,10000,.02,.
     1,5
1030 DATA 3,60000000,10000,.01,.
     1,5
1040 DATA 4,60000000,10000,.01,.
     1,5
1050 DATA 5,60000000,10000,.01,.
     1,5
1090 DATA 0,0,0,0,0,0
1100 DATA 1,4,20000,.02,.05,10,2
     ,100
1110 DATA 2,4,10000,.02,.05,10,2
     ,100
1120 DATA 3,5,10000,.02,.05,10,2
     ,100
1130 DATA 4,5,20000,.02,.05,10,2
     ,100
1140 DATA 5,0,20000,.01,.04,15,1
     ,200

SIMULATIONt,INTERVAL,TABULATIONt
(all mins)?
360,30,60
PRECEEDING WK RAINmm,HYDROGPH CH
ANL no,QMAXm3/s?
100,5,1000
RAINmm/h
?
30
RAINmm/h
?
30
RAINmm/h
?
30
RAINmm/h
?
30
RAINmm/h
?
30
RAINmm/h
?
30
RAINmm/h
?
0
RAINmm/h
?
0
RAINmm/h
?
```

OVERLAND & CHANNEL KINEMATIC FL
OW SIMULATION CATSIM

CHANL Qm3s DEPTHm		
TIME mins		60
1	3.53	.287
2	7.80	.462
3	5.01	.354
4	2.95	.257
5	2.58	.201
TIME mins		120
1	25.12	.931
2	50.92	1.423
3	32.95	1.096
4	33.54	1.108
5	21.21	.710
TIME mins		180
1	76.63	1.819
2	104.12	2.186
3	74.53	1.788
4	173.12	2.965
5	143.32	2.235
TIME mins		240
1	76.79	1.821
2	72.81	1.764
3	56.09	1.508
4	217.33	3.399
5	346.16	3.794
TIME mins		300
1	61.03	1.586
2	54.62	1.484
3	44.38	1.310
4	178.29	3.018
5	299.22	3.477
TIME mins		360
1	46.93	1.355
2	40.99	1.249
3	34.79	1.132
4	139.53	2.605
5	239.00	3.038

Qm3/s- 1000 IN CHANL 5

Thours

Overland flow distance, m.

Average slope over the sub-catchment in the direction of flow towards the channel, m/m

Manning roughnesses of the sub-catchment (note roughness for first catchment is also used for all flood planes)

Steady ultimate infiltration loss in mm/h

The end of the sub-catchment data is identified by typing in a line of six zeros separated by commas.

Data Lines; Channel reaches;

Channel number

Downstream channel number

Length of channel, m

Slope m/m

Manning roughness

Bed width

Channel depth, m

Flood plane width per m depth

The last channel must be the lowest (downstream) channel which is identified by its downstream channel no. (the second item of data in the line) being zero.

The program commences printing a table of channel nos, flow rates in m^3/s and water depths in m, at each successive time interval. A plot of the hydrograph at the designated node appears on the screen simultaneously and this hydrograph is subsequently plotted on paper.

Infiltration and Seepage

In the previous example, excess rainfall was routed overland. That is the user has to insert an infiltration rate for each sub-catchment in the data. In fact losses include an initial abstraction and then a time decreasing infiltration. The Horton (1933) equation suggests an exponentially decreasing loss, whereas the Green-Ampt (1911) equation indicates a less rapid decrease in absorption. The latter equation is based on a simplistic model of soil pores and the decrease in infiltration is based on saturation of the soil pores. Either of these equations could be programmed readily and the indices made functions of preceding rainfall or moisture conditions.

A portion of the infiltration reaches the water table. The water below this level flows laterally under a hydraulic gradient. The deeper the groundwater the higher the flow rate, which can however be exceedingly slow and may not contribute to the hydrograph due to a storm except as a fairly steady basic flow. Where the water table is high however, or if there is a perched water table or high rock level, inter-face flow may occur during or soon after a storm. That is seepage emerges on the surface or into streams shortly after a storm commences and this must therefore be included in the model. Ground water flow can be modelled using the kinematic equations too.

REAL-TIME MODELLING

A model such as that described above was employed on a real-time basis to predict inflows into a reservoir feeding a hydro-electric station (Stephenson, 1986). Rainfall signals from tipping-bucket gauges distri-buted over the 5000 square kilometre catchment were telemetered to a central processing unit linked to two micro computers. One computer processed the data, filed it on floppy discs and at the same time gave a visual display and printout of rainfall over the last hour, and summaries of totals for preceding week etc. The other computer had the catchment model which predicted flow rate into a reservoir. Both storm runoff on a short term basis and low flows over a longer time span were predicted enabling the reservoir to be operated to optimize hydro-power generation.

The hardware and data gathering system were thus low cost but compatible with the accuracy which could be expected. The computer program, based on the kinematic equations and 15 sub-catchments was also at a level matching the data and accuracy which could be expected.

The program is also able to predict ahead the flows based on alter-native assumed rainfall rates. The system lag was 12 to 24 hours, which was generally sufficient to operate gates, but not for conservation of water over the dry season.

236

REFERENCES

Alley, W.M. and Veenhuis, J.E., 1979. Determination of basin characteristics for an urban distributed routing, rainfall–runoff model, in
Proceedings, Stormwater Management Model (SWMM) Users Group Meeting,
pp.1–27.

Crawford, N.H. and Linsley, R.K., 1966. Digital simulation in hydrology:
Stanford Watershed Model IV. Technical Report 39, Civil Engineering
Department, Stanford University, California.

Dawdy, D.R., Lichty, R.W. and Bergman, J.M., 1972. A rainfall runoff
simulation model for estimation of flood peaks for small drainage
basins. U.S. Geological Survey Professional Paper 506–B.

Dawdy, D.R., Schaake, J.C., Jr. and Alley, W.M., 1978. User's guide for
distributed routing rainfall runoff model. U.S. Geological Survey–Water
Resources Investigations 78–90.

Green, W.H. and Ampt, G.A., 1911. Studies of soil physics, I, Flow of air
and water through soils. J. Agric. Science, 4(1), p 1–24.

Holton, H.N., 1961. A concept for infiltration estimates in watershed
engineering. U.S. Dept. of Agriculture, Agriculture Research Service,
ARS–41–45, Washington, D.C.

Holton, H.N. et al. 1975. USDAHL–74 Revised model of watershed hydrology.
U.S. Dept. of Agriculture, Agricultural Research Service, Technical
Bulletin No. 1518, Washington, D.C.

Horton, R.E., 1933. The role of infiltration in the hydrological cycle.
Trans. Am. Geophys. Union., Hydrology papers, p 446–460.

Lapidus, L. and Pinder, G.F., 1982. Numerical solution of partial
differential equations in science and engineering. John Wiley and Sons,
New York, N.Y.

LeClerc, G. and Schaake, J.C., Jr., 1973. Methodology for assessing the
potential impact of urban development on urban runoff and the relative
efficiency of runoff control alternatives. Ralph M. Parsons Laboratory
Report No. 167, Massachusetts Institute of Technology, Cambridge, Mass.

Philip, J.R., 1954. An infiltration equation with physical significance.
Proceedings of the Soil Science Society of America, Vol. 77, pp. 153–157.

Rosenbrock, H.H., 1960. An automatic method of finding the greatest or
least value of a function. Computer Journal, Vol. 3, pp. 175–184.

Ross, B.B., et al. 1978. A model for evaluating the effect of land uses on
flood flows. Bulletin 85, Virginia Water Resources Research Center,
Virginia Polytechnic Institute and State University, Blacksburg,
Virginia.

Ross, B.B., et al. 1982. Model for simulating runoff and erosion in
ungauged watersheds. Bulletin 130, Virginia Water Resources Research
Center, Virginia Polytechnic Institute and State University, Blacksburg,
Virginia.

Shanholtz, V.O. and Lillard, J.H., 1970. A soil water model for two
contrasting tillage systems. Bulletin 38, Virginia Water Resources
Research Center, Virginia Polytechnic Institute and State University,
Blacksburg, Virginia.

Stephenson, D., 1986. Real-time kinematic catchment model for hydro
operation. Proc. ASCE Energy Journal.

CHAPTER 12

GROUNDWATER FLOW

GENERAL COMMENTS

Although the majority of the book analyses surface runoff the same theory is applicable to sub-surface flow. That is kinematic analysis can be used to study a majority of groundwater flow problems that relate to catchment yield and response to storms.

The largest problem is relating the contribution to groundwater flow to infiltration. Although practically all lateral flow underground occurs in the zone beneath the water table, all water permeating in from the surface does not reach the water table. Some is held by capillary forces onto soil particles in the semi saturated zone.

There are also in some situations barriers to flow under the surface. Either obstacles in the path of the flow stop lateral flow, (Harr, 1977) or horizontal impermeable layers create perched water tables which results in more than one lateral flow path underground (Weyman, 1970). The water table can in some instances emerge on the surface, either for the rest of the flow path down to a stream, or to disappear again when a more porous aquifer is reached. The analysis below is therefore somewhat simplistic but demonstrates the principles of kinematic hydrology can be used to estimate groundwater flows. The accuracy of the analysis is more likely to be limited by lack of data on the aquifer than the analytical method. The mechanism of groundwater contribution on slopes was explained by Dunne (1978). Further analysis of the role of sub-surface flow is given by Freeze (1972).

FLOW IN POROUS MEDIA

Whereas for overland flow the Manning equation was found most applicable for estimating flow-depth relationships, flow through ground is generally laminar.

In the flow equation

$$q = \alpha y^m \tag{12.1}$$

for laminar overland flow

$$\alpha = gS/3\nu \tag{12.2}$$

and $m = 3$ where ν is the kinematic viscosity of the fluid, g is gravity and S is the energy gradient, and for turbulent and overland flow

$\alpha = S^{1/2}/N$ for the Manning equation (12.2b)

and $\alpha = CS^{1/2}$ for the Chezy equation. (12.2c)

For flow through porous media the head loss is generally laminar but the general equation is

$$S = av^b \qquad (12.3)$$

where b is 1 for laminar flow, increasing through 1.85 for coarse particles (Ahmed and Sunada, 1969) to 2 for turbulent flow through large rocks (Stephenson, 1979). A general expression is

$$S = Kv^2/gdn^2 \qquad (12.4)$$

where $K = K_1 e\nu/vd + K_2$ (12.5)

here e is the porosity, v the apparent velocity q/h and d a representative grain size. For most aquifers K_2 is negligible so

$$S = K_1\nu v/ged^2 \qquad (12.6)$$

generally an equation of the following form is used

$$S = v/k \qquad (12.7)$$

where k is the permeability,

then q = kSh (12.8)

so $\alpha = kS$ and m = 1 (12.9)

It is assumed the scale of the system is such that surface tension forces can be neglected. Although these may be significant above the water table the lateral flow in this region is usually negligible.

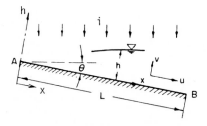

Fig. 12.1 Definition sketch for flow over a sloping plane.

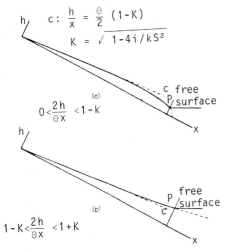

Fig. 12.2 Steady groundwater flow over a sloping plane into a stream

DIFFERENTIAL EQUATIONS IN POROUS MEDIA

Since flow velocities in porous media are as a rule very small, the depth of saturated aquifer can be large to discharge the flow. The slope of the water table may therefore differ from the slope of the impermeable bed and the kinematic dynamic equation may require modification. The diffusion equation is thus often used. In Fig. 12.1 the aquifer is assumed to overlay an impermeable plane. The only outside contribution is assumed to be the infiltration from above, i (m/s). The rate will be assumed constant in the analysis below. Varying aquifer transmissivity and allowance for partial saturation is made later. Owing to the slow velocities, omission of the dynamic terms is even more justified than for overland flow.

The continuity equation becomes

$$\frac{\partial q}{\partial x} + e \frac{\partial h}{\partial t} = i \qquad (12.10)$$

The dynamic equation becomes

$$q = k \left(Sh - h \frac{\partial h}{\partial x} \right)$$

This is termed the Dupuit-Forchheimer equation. The first term on the right hand side gives the Darcy equation and the second term is the correction for water surface gradient, as in the diffusion equation. Elimination of q from the continuity and diffusion equation yields

$$\frac{e}{k} \frac{\partial h}{\partial t} + S \frac{\partial h}{\partial x} = \frac{\partial}{\partial x} \left(y \frac{\partial h}{\partial x} \right) + \frac{i}{k} \qquad (12.11)$$

Henderson and Wooding (1964) solved this equation for certain cases to show the depth of emergence is influenced by the downstream conditions (see Fig. 12.2).

ANALYSIS OF SUBSURFACE FLOW

Freeze (1972) indicated that subsurface flow could only generate stormflow i.e. a relatively short time to peak hydrograph in cases where the soil is very permeable. Where the slope is very steep however, Beven (1981) indicates interflow (flow under and over surface successively) could occur which would accelerate the concentration process.

Beven (1981) extended Henderson and Wooding's (1964) analysis for kinematic subsurface flow using a dimensionless form of the extended Dupuit-Forchheimer equations:

$$\frac{\partial H}{\partial T} = \frac{\partial}{\partial X} \left(H \frac{\partial H}{\partial X} \right) - 2 \frac{\partial H}{\partial X} + \lambda \tag{12.12}$$

where

$$X = x/L \tag{12.13}$$

$$H = 2h/L \tan \theta \tag{12.14}$$

$$T = k \sin \theta \, t/2eL \tag{12.15}$$

$$\lambda = 4i \cos \theta /k \sin^2 \theta \cong 4i/kS_o^2 \tag{12.16}$$

Omitting the diffusion term

$$\frac{\partial H}{\partial T} = -2 \frac{\partial H}{\partial X} + \lambda \tag{12.17}$$

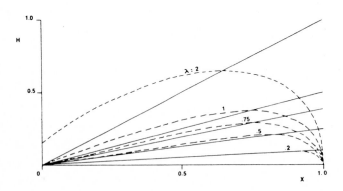

Fig. 12.3 A comparison of steady state water table profiles predicted by the extended Dupuit-Forchheimer (broken lines) and kinematic wave (solid lines) equations for different values of λ. (K. Beven, Water Resources Research, 17, 1422, 1981, Copyright American Geophysical Union).

results in the kinematic equation which may be integrated to give the
rising hydrograph at x = L

$$H = \lambda T \, (T < Tc = 0.5)$$ (12.18)

A comparison of this solution with numerical solutions of the
extended Dupuit–Forchheimer equation was made by Beven (1981) who
used an implicit finite difference method and iterative relaxation solution.
The results are given in Fig. 12.3 and 12.4. They indicate the kinematic
equation holds reasonably for $\lambda < 1$.

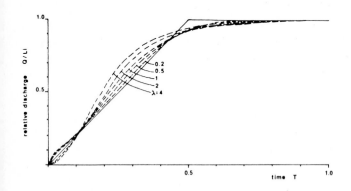

Fig. 12.4   Rising hydrographs predicted by the extended Dupuit–Forchheimer
(broken lines) and kinematic wave (solid line) equations for
different values of λ. (K. Beven, Water Resources Research,
17, 1423, 1981, Copyright American Geophysical Union).

FLOW IN UNSATURATED ZONE

Beven (1982) extended his analysis of the saturated zone to allow
for vertical flow in the unsaturated zone above the water table. He also
accounted for variations in porosity and hydraulic conductivity with
depth.

Starting from Campbell's (1974) model the hydraulic conductivity
is based on the equations

$$k = \frac{K(\theta)}{K_s} = S^{2B+3}$$ (12.19)

$$S = \frac{\theta}{\theta_s} = \left(\frac{\psi_b}{\psi}\right)^{1/B}$$ (12.20)

where k is the relative hydraulic conductivity, $K(\theta)$ is hydraulic
conductivity at moisture content θ, subscript s refers to saturation
conditions, S is relative saturation, ψ is capillary tension, ψ_b is tension

at air entry and B is a pore size parameter, which varies from about 4 for sands to 11 for clays. It is assumed that

$$K_s = K_* (D - Z)^n = K_* h^n \tag{12.21}$$

$$\Theta = \Theta_* (D - Z)^m = \Theta_* h^m \tag{12.22}$$

where Z is depth below soil surface perpendicular to the slope, D is the constant depth of soil, K_*, Θ_*, n and m are constants ($n \cong 2m$). Starting at $t = o$ with constant initial capillary tension ψo and $h = o$,

$$\tag{12.23}$$

Darcy's law for unsaturated flow is

$$q_z = -K(\Theta) \frac{\partial \psi}{\partial z} + K(\Theta) \cos \phi \tag{12.24}$$

where q_z is the volume flow perpendicular to the slope and ϕ is the slope angle to the horizontal.

Assuming water due to infiltration moves down with an interface parallel to the slope, at a rate

$$\frac{dz}{dt} = \frac{i}{\Theta_w(z) - \Theta(z, t=0)} \tag{12.25}$$

where $\Theta_w(z)$ is the water content at which $K(\Theta, z) \cos \phi = i$

Then $\Theta_w(z) = \Theta_* \left(\frac{1}{K_* \cos \phi}\right)^{1/(2B+3)} (D-Z)^{m-n/(2B+3)} \tag{12.26}$

$$= a(D-Z)^b = ah^b \tag{12.27}$$

where $a = \Theta_* \left(\frac{i}{K_* \cos \alpha}\right)^{1/(2B+3)}$, $b = m - \frac{n}{2B+3} \tag{12.28}$

The above equations apply until the wetted front reaches a point at which $i = K_s(z) \cos \phi$

Below this depth $h_w = (i/K_* \cos \phi)^{1/n} \tag{12.29}$

$$\frac{dz}{dt} = \frac{i}{\Theta_s(z) - \Theta(z, t=o)} \tag{12.30}$$

Integrating the two equations for z gives the time t_{uz} at which the wetting front reaches the bottom of the profile.

$$t_{uz} = \frac{1}{i} \left\{ \frac{a}{1+b} (D^{1+b} - h_w^{1+b}) + \frac{\Theta_*}{1+m} (h_w^{1+m} - \left(\frac{\Theta_*}{1+m}\right)^{1/B} D^{1+m}) \right\} \tag{12.31}$$

FLOW IN NON-HOMOGENEOUS SATURATED ZONE

If it can be assumed the water table is parallel to the impermeable bed, the hydraulic gradient is $\sin \phi$ and

$$q_x = \int_o^h K_s(h) \sin \phi \, dz = \frac{K_* \sin \phi h^{n+1}}{n+1} \tag{12.32}$$

The kinematic wave equation thus becomes

$$e(h) \frac{\partial h}{\partial t} = -K_* \sin \phi \, h^n \frac{\partial h}{\partial x} + i \qquad (12.33)$$

where $t > t_{uz}$. If input continues until $t > t_{uz}$ the steady state unsaturated profile has porosity $e(h) = \Theta_s(h) - \Theta_w(h)$ $\qquad (12.34)$

$$= 0 \; ; \quad h \le h_w \qquad (12.35)$$

$$= \Theta_* h^m - a h^b \; ; \; h > h_w \qquad (12.36)$$

Substituting into the previous equation gives

$$\frac{\partial h}{\partial t} = -(\frac{K_* \sin \phi \, h^w}{\Theta_* h^m - a h^b}) \frac{\partial h}{\partial t} + \frac{i}{\Theta_* h^m - a h^b} \qquad (12.37)$$

Integrating the characteristic equation yields

$$t = t_{uz} + \frac{1}{i} \{ \frac{\Theta_*}{1+m} (h^{1+m} - h_w^{1+m}) - \frac{a}{1+b} (h^{1+b} - h_w^{1+b}) \} \qquad (12.38)$$

When the characteristic from the top of the catchment reaches the outlet, a steady state is established and

$$h = (\frac{(n+1)ix}{K_* \sin \phi})^{1/1+n} \qquad (12.39)$$

and the time of concentration is

$$t_c = t_{uz} + \frac{1}{i} \{ \frac{\Theta_*}{1+m} (h_L^{1+m} - h_w^{1+m})$$

$$- \frac{a}{1+b} (h_L^{1+b} - h_w^{1+b}) \} \qquad (12.40)$$

A solution for the falling limb of the hydrograph is also possible if it can be assumed a drying front descends uniformly once rainfall stops.

After surface infiltration ceases the drying front continues to fall at a rate

$$\frac{dz}{dt} = \frac{i}{\Theta_w(z) - \Theta_d(z)} \qquad (12.41)$$

The wetted profile is given by

$$\Theta_w(z) = \Theta_* (\frac{1}{K_* \cos \phi})^{1/(2B+3)} (D-Z)^{m-n/(2B+3)} \qquad (12.42)$$

and when drying

$$\Theta_d(z) = \Theta_* h^m (\frac{\psi_b}{\psi_d})^{1/B} \qquad (12.43)$$

Integrating the two equations yields the time after cessation of rain until the drying front reaches the water table

$$t_d = t_r + \frac{1}{i} \{ \frac{a}{1+b} (D^{1+b} - h^{1+b}) - \frac{\Theta_*}{1+m} (\frac{\psi_b}{\psi_d})^{1/B} (D^{1+m} - h^{1+m}) \} \qquad (2.44)$$

where $z = o$ at $t = t_r$

Fig. 12.5 depicts the equilibrium hydrographs for different rainfall rates.

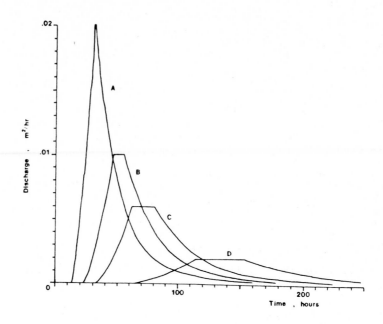

Fig. 12.5 Equilibrium hydrographs resulting from rainfalls of different intensities and duration. Model parameters are as in Table 12.1 with L_e = 10 m. A is 0.002 m/hr for 30 hours; B 0.001 m/hr for 50 hours: C 0.0006 m/hr for 70 hours and D 0.0002 m/hr for 120 hours. (K. Beven, Water Resources Research, 18, 1631, 1982, Copyright American Geophysical Union).

TABLE 12.1 Model parameters with values used in example calculations

Parameter	Symbol	Value
Soil depth	D	0.6m
Effective slope length	L_e	10.0, 14.0m
Hydraulic conductivity parameter	K	2.0057 m/hr
Hydraulic conductivity parameter	e	2.73
Porosity parameter	θ	0.8035
Porosity parameter	m	1.135
Soil moisture characteristic parameter	ψ_b	10.0 cm
Soil moisture characteristic parameter	B	5.0 cm
Initial moisture tension	ψ_0	500.0 cm
Drying moisture tension	ψ_d	300.0 cm
Slope angle	ϕ	15°

REFERENCES

Ahmed, N. and Sunada, D.K., 1969. Nonlinear flow in porous media. Proc. ASCE, J. Hydr. Div. HY6, Nov. pp 1847-1857.

Beven, K. 1981. Kinematic subsurface stormflow. Water Resources Research 17(5), pp 1419-1424.

Beven, K. 1982. On subsurface stormflow: Predictions with simple kinematic theory for saturated and unsaturated flows. Water Resources Research, 18(6), pp 1627-1633.

Campbell, G.S. 1974. A simple method for determining unsaturated conductivity from moisture retention data. Soil Sci., 117, pp 311-314.

Dunne, T. 1978. Field studies of hillslope flow processes. In Hillslope Hydrology, Ed. M.J. Kirby, John Wiley, N.Y.

Freeze, R.A. 1972. Role of subsurface flow in generating surface runoff, 2, Upstream source area. Water Resour. Res. 8, pp 1272-1283.

Harr, R.D. 1977. Water flux in soil and subsoil on a steep forrested slope, J. Hydrol. 33, pp 37-58.

Henderson, F.M. and Wooding, R.A. 1964. Overland flow and groundwater flow from a steady rainfall of finite duration. J. Geophys. Research, 69 (8), pp 1531-1540.

Stephenson, D. 1979. Rockfill in Hydraulic Engineering, Elsevier, 215 p.

Weyman, D.R. 1970. Throughflow on hillslopes and its relation to the stream hydrograph. Bull. Int. Assoc. Sci. Hydrol. 15 (2), pp 25-33.

AUTHOR INDEX

INDEX